T0215465

Climate Change Justice and Global Resource Commons

This book examines the multiple scales at which the inequities of climate change are borne out.

Shangrila Joshi engages in a multi-scalar analysis of the myriad ways in which various resource commons – predominantly atmosphere and forests – are implicated in climate governance, with a consistent emphasis throughout on the justice implications for disenfranchised communities. The book starts with an analysis of North-South inequities in responsibility, vulnerability, and capability, as evidenced in global climate treaty negotiations from Rio to Paris. It then moves on to examine the ways in which structural inequalities are built into the conceptualization and operationalization of various neoliberal climate solutions such as Reducing Emissions from Deforestation and Forest Degradation (REDD+) and the Clean Development Mechanism (CDM). Drawing on qualitative interviews conducted in Delhi, Kathmandu, and the Terai region of Nepal, participant observation at the Climate Conference in Copenhagen (COP-15), and textual analysis of official documents, the book articulates a geography of climate justice, considering how ideas of injustice pertaining to colonialism, race, Indigeneity, caste, gender, and global inequality intersect with the politics of scale.

This book will be of great interest to students and scholars of environmental justice, climate justice, climate policy, political ecology, and South Asian studies.

Shangrila Joshi is a member of the faculty at The Evergreen State College, USA.

Routledge Studies in Environmental Justice

This series is theoretically and geographically broad in scope, seeking to explore the emerging debates, controversies and practical solutions within Environmental Justice, from around the globe. It offers cutting-edge perspectives at both a local and global scale, engaging with topics such as climate justice, water governance, air pollution, waste management, environmental crime, and the various intersections of the field with related disciplines.

The *Routledge Studies in Environmental Justice* series welcomes submissions that combine strong academic theory with practical applications, and as such is relevant to a global readership of students, researchers, policy-makers, practitioners and activists.

Ecosocialism and Climate Justice
An Ecological Neo-Gramscian Analysis
Eve Croeser

Climate Change Justice and Global Resource Commons
Local and Global Postcolonial Political Ecologies
Shangrila Joshi

For more information about this series, please visit: www.routledge.com/Routledge-Studies-in-Environmental-Justice/book-series/EJS

Climate Change Justice and Global Resource Commons

Local and Global Postcolonial Political Ecologies

Shangrila Joshi

Routledge
Taylor & Francis Group
LONDON AND NEW YORK

earthscan
from Routledge

First published 2021
by Routledge
2 Park Square, Milton Park, Abingdon, Oxon OX14 4RN

and by Routledge
52 Vanderbilt Avenue, New York, NY 10017

Routledge is an imprint of the Taylor & Francis Group, an informa business

British Library Cataloguing-in-Publication Data
A catalogue record for this book is available from the British Library

Library of Congress Cataloging-in-Publication Data
Names: Joshi, Shangrila, author.
Title: Climate change justice and global resource commons : local and
    global postcolonial political ecologies / Shangrila Joshi.
Description: Abingdon, Oxon ; New York, NY : Routledge, 2021. | Series:
    Routledge studies in environmental justice | Includes bibliographical
    references and index.
Identifiers: LCCN 2020046789 (print) | LCCN 2020046790 (ebook) |
    ISBN 9780367364557 (hardback) | ISBN 9780429346231 (ebook)
Subjects: LCSH: Environmental justice—Developing countries. | Political
    ecology—Developing countries. | Postcolonialism—Developing
    countries. | Global commons—Developing countries. | Climate change
    mitigation—Developing countries. | Environmental policy—Developing
    countries. | Developing countries—Environmental conditions.
Classification: LCC GE240.D44 J67 2021 (print) | LCC GE240.D44
    (ebook) | DDC 363.738/74091724—dc23
LC record available at https://lccn.loc.gov/2020046789
LC ebook record available at https://lccn.loc.gov/2020046790

ISBN: 978-0-367-36455-7 (hbk)
ISBN: 978-0-367-75129-6 (pbk)
ISBN: 978-0-429-34623-1 (ebk)

Typeset in Bembo
by Apex CoVantage, LLC

For Dai Ba and Didi Ma

"Don't be trapped in someone else's Single Story."
and
"Be indigestible in the belly of the beast."

(Dr. W. Joye Hardiman, at the 2020 Juneteenth
celebration at Tacoma Evergreen)

Contents

Figures

Preface

The inspiration for writing this book comes from the challenge I have felt in selecting an appropriate book on the complexities of the climate justice dilemma for the students I teach at The Evergreen State College. A public liberal arts college in the Pacific Northwest, Evergreen, is renowned for its progressive outlook and credentials for producing graduates who are passionate about changing the world for the better by getting to the roots of the various problems that plague the contemporary world.

Students who come to Evergreen are generally passionate about environmental stewardship and about social justice, but among these progressive communities, there tends to be an obliviousness about the world outside the United States that I find troubling. My goal as a teacher is to challenge students to think about environmental and climate justice in ways that decenter the United States. It is not that these aspirations are not necessary for communities within the United States but rather that one cannot develop a sufficient understanding of the complexities of climate change and justice while limiting oneself to the US bubble. The question of the historical responsibility of this country in contributing disproportionately to the earth's warming over the past century and more is, I believe, of relevance to all who reside here. Even those who have been disadvantaged and marginalized within the United States enjoy the relative privileges of being located within the world's hegemonic power. What to make of the obligations the rest of the world projects on the country that also has a historical responsibility to those disenfranchised and exploited within its borders? Should the relationship between the two sets of claimants for reparations from the United States be one of competition? Of solidarity? How to think of the culture of hubris and exceptionalism that marks the image of the United States abroad? Is it limited to privileged groups or more widespread in the culture?

These are some of the thorny questions this book will seek to articulate and start to address. It will do so by connecting a number of disparate but connected areas of academic inquiry that have not yet had a chance to cross-pollinate in the space of one volume. As such, notions of scale, environmental and climate justice, political ecology, critical geopolitics, global environmental politics, Critical Race Theory, postcolonial theory, and decolonial thought will all be put in conversation with one another in order to unpack the knotty dilemmas of the climate justice puzzle.

I write from the positionality of a woman of color in the United States who finds the emerging BIPOC (Black, Indigenous, and People of Color) discourse important and validating but US-centric in a way that is not adequate to the task of adequately grappling with the crisis of climate injustice. To undo the harms done by settler colonialism, slavery, and the vestiges of racism inherent in both historical injustices is important in its own right. It is also important to recognize that these oppressive structures contribute to creating vulnerabilities to climate impacts in the affected communities within the United States. But a climate justice agenda that centers racial, Indigenous, gendered, and/or economic disenfranchisement in the United States while leaving out consideration of how the rest of the world is affected by climate change and the role the United States plays in the world is problematic. From my positionality of a person born and raised in South Asia, but who has a voice in US academia, my hope is to emphasize the global dimension of climate injustice and to work against the invisibilization of the Global South. Towards this effort I draw not only on my scholarly research and analysis shaped by the fields of study of political ecology and environmental justice but from my lived experience with this positionality in the world and in my profession of teaching.

Acknowledgments

This book would not have been possible without the institutions and individuals that have supported my work. I wish to thank the University of Oregon, The Evergreen State College, and the Tokyo Foundation for supporting my research and writing endeavors over the years. Many thanks to the Routledge team for seeing this project to completion. The numerous organizations, communities, and individuals in India and Nepal where I conducted research deserve my deepest gratitude. Sharing perspectives on interview questions, not knowing if there would be any tangible benefits, requires trust. I am grateful to all the individuals interviewed for that trust. I hope I did justice to your sentiments and created adequate space in this book for your voices. Gratitude is also due to my teachers, colleagues, and students past, present, and future, for inspiring and/or challenging me to broaden my worldview and to ask new kinds of questions.

And, I wish to thank members of my immediate and extended family for being a consistent source of support. Without my parents' blessings, none of this work would have been accomplished, nor without the endless contributions of my spouse who carried the lion's share of reproductive labor in our household while I dedicated myself to the work of writing. As always, any of this work is done with borrowed time from fulfilling the responsibility of raising our child. In this respect, I cannot thank my teenage son enough for displaying understanding and maturity towards my work obligations, particularly during the pandemic, when we spent months sequestered in our house. I am grateful for our badminton games, crossword puzzles, and card games, as well as for working together to create the index for this book. I am grateful for daily check-ins with my mother, and for my brother's critical input on early drafts.

Abbreviations

AOSIS	Alliance of Small Island States
BIPOC	Black, Indigenous, and People of Color
CBDRRC	Common but Differentiated Responsibilities and Respective Capabilities
CF	Community Forestry
CFUG	Community Forest User Group
COP	Conference of Parties
CSE	Centre for Science and Environment
DDT	Dichloro-diphenyl-trichloroethane
DFO	District/Division Forest Office
ERPA	Emission Reductions Payment Agreement
EJ	Environmental Justice
FAB	Fair Ambitious and Binding
FCPF	Forest Carbon Partnership Facility
FECOFUN	Federation of Community Forestry Users-Nepal
FPIC	Free, Prior, and Informed Consent
GESI	Gender Equality and Social Inclusion
GHG	Greenhouse Gas
G-77	Group of 77
GoI	Government of India
GoN	Government of Nepal
ICIMOD	Integrated Centre for International Mountain Development
IEN	Indigenous Environmental Network
ILO	International Labour Organization
INDC	Intended Nationally Determined Contributions
IPCC	Intergovernmental Panel on Climate Change
IYCN	Indian Youth Climate Network
JNU	Jawaharlal Nehru University
LGBT	Lesbian, Gay, Bisexual, and Transgender
LPG	Liquefied Petroleum Gas
LRP	Local Resource Person
MoEF	Ministry of Environment and Forests
MoEWRI	Ministry of Energy, Water Resources and Irrigation

MoFE	Ministry of Forests and Environment
MoSTE	Ministry of Science, Technology, and Environment
MRV	Monitoring, Reporting, and Verification
MtCO2e	Metric tons of Carbon Dioxide Equivalent
NEFIN	Nepal Federation of Indigenous Nationalities
NGO	Non-Governmental Organization
NIEO	New International Economic Order
NRM	Natural Resource Management
PDD	Proposal Design and Development
PMCCC	Prime Minister's Council on Climate Change
RBF	Results-Based Finance
REDD	Reducing Emissions from Deforestation and Forest Degradation
SFM	Sustainable Forest Management
TERI	The Energy and Resources Institute
UN	United Nations
UNCED	United Nations Conference on Environment and Development
UNCHE	United Nations Conference on the Human Environment
UNFCCC	United Nations Framework Convention on Climate Change
US	United States
VDC	Village Development Committee
WB	World Bank
WWF	Worldwide Fund for Nature

1 Introducing climate change as a global commons problem

The year 2020 was destined to be a year of reckoning, even before the pandemic hit, even before the wrongful deaths of George Floyd and Breonna Taylor precipitated a reinvigorated Black Lives Matter (BLM) movement in the United States, with reverberations all over the globe, bringing to public discourse the notion that White supremacy is an unacceptable way of being. The year also marks the end of the Kyoto Protocol, and the official transition from this global treaty to the Paris Agreement, during a year that gave us blazing forest fires from Australia to the Arctic Circle, locust outbreaks in Africa and South Asia, Cyclone Amphan in the Bay of Bengal, Typhoon Goni in the Philippines, and Hurricane Eta in Central America among other calamities, creating a climate in which it has become evermore incredulous for a reasonable person to turn their backs on the seriousness of the climate crisis. As such, this is an opportune moment to reflect on this change – this moment of transition from the Kyoto Protocol to the Paris agreement – and what it signifies for global climate justice, for US exceptionalism in climate negotiations, and beyond. The confluence of multiple crises in 2020 enables us to see connections, recognize patterns, and to potentially identify ways to bounce forward from these and other global crises.

Unlike the problem of ozone thinning, the problem of global climate change has not been adequately addressed at the international level. A big deterrent has been the foot-dragging of the highest cumulative emitting country in the world – the United States – that infamously remained the only Western country to never have ratified the Kyoto Protocol, the reigning global treaty on climate change that is set to expire at the end of 2020. It will be replaced by the Paris Agreement, as was decided during the 2015 Paris Climate Conference. The Trump administration's move to withdraw from even the Paris Agreement – which is a weaker treaty than the Kyoto Protocol – was widely condemned. But it is important to understand that even in the pre-Trump era, there was much that was wanting in the United States' demeanor towards the international process of tackling an environmental problem that is truly global in scope. The US role in international treaty negotiations has been marked by unilateralism, exceptionalism, marketization, obstructionism, abandonment, and most recently by a shameless disregard for treating the universe as a global commons.

Following the rise of the environmental movement in the United States in the 1960s, the first major international environmental meeting in Stockholm, the United Nations Conference on the Human Environment (UNCHE), started the process where powerful industrialized countries could pressurize newly independent nation-states that were emerging from a centuries-long history of colonial rule to accept direction in how they should pursue development, due to the consequences this pursuit may have on the natural environment – thanks to the newly developed environmental awareness in the West. The United States exercised unilateralism with its trade partners in South America, such as in the case of dolphin-safe tuna, imposing these newly acquired sensibilities to place constraints on resource use elsewhere. Not surprisingly, Global South countries balked at the perceived green neocolonialism on part of Global North countries (Castro 1972). By the time the UN Conference on Environment and Development (UNCED) met in Rio de Janeiro in 1992, two decades after Stockholm, the environmental agenda – often pushed by the Global North – was inextricably linked with the development agenda – seemingly a priority of the Global South. The UN Framework Convention on Climate Change (UNFCCC) emerged from the Rio Earth Summit, and the notion of common but differentiated responsibilities and respective capabilities (CBDRRC) was institutionalized as a principle deemed fair given the differential positioning of recognized nation-states in global political and economic power, privilege, and responsibility. The CBDRRC principle was in essence an international norm negotiated by the representatives of the international community in an effort to help create a fair global climate treaty (Miller 1995; Williams 2005). The United States famously disagreed to abide by the obligations that adherence to this principle would place on its industry and economy, vis-à-vis those of China and India that were deemed as 'emerging economies'. George H. W. Bush is known to have made the outrageous claim in 1992 that "the American lifestyle is not open to negotiation" during the Rio Summit (Roberts and Parks 2007, 3). While Bush eventually agreed to differential obligations for the United States vis-à-vis India and China during negotiations in the lead up to the Montreal Protocol to address ozone thinning – and this treaty has gone on to effectively curtail the severity of ozone thinning – the same has not transpired for the problem of anthropogenic climate change that has fossil fuel-based industrialization as a key driver, which also happens to be central to the way of life Bush was unwilling to negotiate.

Ironically, the United States' refusal to ratify the Kyoto Protocol follows a prolonged and deliberate period of negotiations where the United States successfully lobbied to include a cap-and-trade market-based mechanism as a key element of the treaty, weakening it considerably by creating the conditions for the commodification of the atmosphere, even as the inclusion of market-based principles for flexibility was meant to be a precondition for the United States to sign (Ervine 2018; Narain 2019). As annual negotiations for the post-Kyoto period continued, the United States engaged in a deliberate process of breaking up Global South unity in climate negotiations, promoting the strategy of challenging the validity of the non-Annex I category of countries (Vidal 2010). Paradoxically, on this point, the United States' position appears uncomfortably aligned with those of critical scholars who challenge the validity of the Global South due to its perceived departure from an earlier, more radical agenda of non-alignment, which led to the creation of the Third World coalition (Barnett 2007; Berger 2004). In the United States, the country of my residence since 2002, there are vibrant and robust grassroots movements that are pushing back against the foundations of the fossil fuel empire within the country. But despite the zeal to push back against the atrocities of the state and to push for action at many local sites, there appears a reluctance to acknowledge the importance also of the international site of action, where representatives of nation-states gather to prepare the rules and norms that would be agreed for the common good. Inaction at this scale is keenly felt by those in the Global South who have been at the frontlines of climate impacts, as this statement from a young Bangladeshi man affected by Cyclone Amphan indicates:

> Sometimes I feel that the international community has also abandoned us. They have left us to die slowly. The West is largely responsible for global warming, which has resulted in higher sea levels. We are the ones who have to suffer because of it. Still, we don't see any international initiative for our protection.
>
> (S. M. Shahin Alam, cited in Islam 2020)

In Chapters Two and Three I explain in more detail the politics of treaty-making at the site of the international, but here I want to argue why the treaty-making process is an important space to work towards the safeguarding of the global commons. The United States unceremoniously withdrew from the Paris Agreement, which adopted the apolitical strategy of encouraging all signatory countries to declare voluntary mitigation targets which would be gradually strengthened, even as it had already been weakened to accommodate the United States' preferences (Narain 2019). The abandonment of the Paris Agreement by the United States is not so much the problem as the movement from the Kyoto Protocol to the Paris Agreement, which I suggest is the movement from treating the atmosphere as a commons – where members agree to abide by rules created for members by members based on a process of deliberation – to treating the atmosphere as an open-access resource, protected only to the extent that individual members' conscience allows. The Kyoto Protocol attempted to constrain global emissions with binding commitments for the international community based on principles of equity; the Paris Agreement embodies a clear departure from North-South equity. Commons scholar Elinor Ostrom (1990, 1999) has documented the existence of common pool resource governance structures in various locations around the world, but scaling up commons governance for global issues such as climate change has proved to be challenging. Nevertheless, the atmosphere is a global resource, whether we treat it as a commons or as an open-access resource, and whether we call it a resource or not – some are put off by the utilitarian feel of the term – even as by nature humans *use* various elements of non-human nature to survive. The significance of treating it as a commons has in my view become more pronounced, given the recent move made by the Trump administration to lay claim to the resources on the moon, Mars, and other celestial bodies, in what would be the ultimate 'spatial fix' (Harvey 2018, 133).

> Americans should have the right to engage in commercial exploration, recovery, and use of resources in outer space, consistent with applicable law. Outer space is a legally and physically unique domain of human activity, and the United States does not view it as a global commons.
>
> (Executive Order 13914 of April 6, 2020, 20381)

The tragedy of the atmospheric commons

The complexities of the global climate crisis are immense. One key source of complexity comes from the transboundary nature of the global greenhouse effect, owing to the diffuse nature of the atmosphere that serves as a sink for greenhouse gases (GHGs). The resultant warming of the planet in turn has unpredictable consequences for the water cycle and its effects on human and non-human assemblages worldwide. The fallout from the impacts are many, widespread, and unpredictable – although the frequency and intensity of climatic events such as cyclones and droughts are known to rise with passing years. The sources of GHGs – carbon dioxide, methane, and others – are many and widespread as well. The combination of the irreducibility of the sources, sink, and consequences of the global greenhouse effect to a particular nation-state or region of the world renders this a classic global scale problem that could not be successfully resolved by mitigation with stringent action being taken within one country's borders.

Since at least 1992, the international community has mobilized to address the emerging global warming issue by formulating a binding treaty as had been the norm for other transboundary and global atmospheric crises such as acid rain and ozone thinning, respectively (Skjaerseth 2012; Wettestad 2012). For each issue, representatives of nation-states gathered at UN-facilitated conferences of parties to negotiate a treaty that would be agreeable to relevant parties. Over two decades of negotiations, a consensus-based binding global treaty was barely mustered in the form of the Kyoto Protocol, which came into effect officially in 2005 after the Russian Duma ratified the treaty in 2004, thus reaching the minimum number of countries accounting for more than a half of global emissions (Roberts and Parks 2007). This was the closest the international community came to devising a set of rules and norms for the atmosphere as if it were a traditional resource commons; the action and behavior of the United States was like that of the classic free-rider.

Conjuring up the idea of the atmospheric commons necessitates me to address here two key works on the commons. The phrase 'the tragedy of the commons' was popularized by Garrett Hardin, who published an article by that name in *Science* in 1968. Based on his unfounded assertion that the commons was an open-access resource – which he later amended (Hardin 1994) – his original argument was that a resource commons necessarily meets a tragic demise due to the inherently selfish nature of humans, who seek to maximize personal gain at the cost of eventual communal loss. Further, he postulated that top-down enclosures of the commons, reducing the global population, and the privatization of the commons were the most effective ways to prevent a tragedy of the commons. Fewer people would mean fewer claims on the commons. Privatization would give each individual property owner an incentive to protect the piece of the commons they now owned. If the question of justice should arise in operationalizing his preferred solution of privatization, he reasoned,

> [a]n alternative to the commons need not be perfectly just to be preferable. . . . We must admit that our legal system of private property plus inheritance is unjust – but we put up with it because we are not convinced, at the moment, that anyone has invented a better system. . . . Injustice is preferable to total ruin.
>
> (Hardin 1968, 1247)

Ostrom's rigorous counterarguments to Hardin's (1968) thesis eventually compelled him to publish his revisionist take in 1994. Ostrom pointed out that in the real world – as opposed to in the hypothetical scenario of the pasture conjured up by Hardin – the traditional resource commons were not always treated as an open-access resource where a person's individual conscience was the only thing restraining them from overuse to their own detriment, due to the gradual elimination of conscientious behaviors in a social Darwinist interpretation of the theory of the 'survival of the fittest'. While Ostrom was careful not to generalize, she presented empirical research based on communities around the world, leading to her claim that in many resource-dependent communities, the resource commons did not exhibit the characteristics laid out by Hardin (1968). In fact, in the most remarkable examples, resource use by well-defined communities emulated traditional egalitarian and democratic governance structures put in place for the continued vitality of the resource in question (Ostrom 1999). Ostrom's (1990) work deduced that the best common pool resource governance structures tend to have the following attributes:

1   A clearly defined group that is capable of excluding non-members.
2   Rules governing equitable distribution of benefits and burdens that are suitable for the local context.
3   Those affected by the rules can participate in modifying the rules through collective action.

4   Community members exercise autonomy in rule-making; community self-governance is recognized by higher-level authorities.
5   Rules are enforced and monitored by community-appointed individuals with accountability to the community.
6   Violators of community-determined rules are sanctioned.
7   Mechanisms available for conflict resolution are low cost and easy to access.
8   Larger common-pool resources employ a tiered organizational structure to facilitate governance.

Despite evidence to the contrary, Hardin's thesis has continued to dominate mainstream environmental problem-solving approaches. For example, national parks and protected areas created to prevent biodiversity and habitat loss for endangered species are classic enclosures of the commons where a centralized authority, such as the government of a nation-state, assumes governance power over the resource. Ordinary citizens living in the vicinity of the protected area – who may be descendants of those who were ousted during the creation of the park, in many cases Indigenous inhabitants of the land – have little to no say over decision-making choices about how the space and resources therein should be managed. The second of Hardin's preferred solutions – privatization of the commons – is currently reflected in the plethora of carbon-trading initiatives that enable individuals, businesses, and other entities to buy and sell carbon credits (Ervine 2018). Carbon trading facilitates the commodification of the atmosphere's ability to absorb GHGs, by allowing those who have used less than their designated share of this ability to sell their carbon credits accrued. Despite counterarguments from scholars and activists, these top-down and neoliberal solutions proliferate, in large part because they don't require the status quo to change. They continue to concentrate power in the hands of either centralized authorities or wealthy property owners. Sometimes the two solutions work in tandem, exacerbating the impact of each – enclosure as well as privatization of the commons.

In this book, I argue that the climate crisis warrants a recalibration of this paradigm and an embracing of the practice of the commons. To do so, the task is not to reinvent the wheel but rather to learn from existing practices – in places where the modernization of resource governance has not yet successfully ousted traditional practices, and in places where deliberate commons management policies have been ushered in. Chapters Five, Six, and Seven discuss this further, drawing on my research in Nepal. A comprehensive study of commons is beyond the scope of this volume, but what I propose will hopefully generate interest in studying this further. I suggest that there is a way to see the commons at disparate local contexts to be connected to the ultimate global atmospheric commons. In each of these contexts, we can choose to treat sources and sinks as a commons or as an open access resource. While research on the commons in local contexts is more common – such as in the context of forests, irrigation systems, and fisheries – scaling up commons practices for international governance has proved to be challenging. Among other reasons, clear delineation of boundaries between insiders and outsiders and the question of how to create meaningful sanctions for bad behavior, particularly when members of the community are nation-states, have been deemed insurmountable. Yet, we managed to treat the atmosphere as a commons when faced with the predicament of ozone thinning (Gibson-Graham, Cameron, and Healy 2013).

The UNFCCC and the annual UN meetings on climate change since 1992 – the various Conferences of Parties (COPs) – have served as a deliberative space to facilitate collaborative decision-making of this global-scale problem, albeit with significant limitations in representation and the disproportionate power of free riders. There have been numerous critiques of the UN deliberation process due to these limitations – and they need to be heeded if the possibility of common pool resource governance of the atmosphere is to be seriously taken. In addition to the limits of inadequate representation of those that are marginalized and disenfranchised by the state – whether Indigenous groups, ethnic minorities, women, youth, LGBT, or other groups that tend not to be proportionally represented by state officials – the UNFCCC will have to account for the challenges posed by the multi-scalar nature of the sites of climate action, as well as the complications created by the politics of scale at multiple sites of action and discourse, that can confound the conceptualization and pursuit of climate justice. Unlike the sources of ozone-depleting chemicals, the sources of GHGs are many and diffuse, and this further complicates identification and regulation of the members of the community whose overproduction of GHGs is to be curtailed. Therefore, scaling up common pool resource governance for the atmosphere is more complicated than it would seem. Yet the alternative encourages all to be free riders, to everyone else's detriment eventually, and in the short term, to the immediate and worst detriment of those who have contributed the least to creating the problem of climate change. Instead of disavowing UN treaties, therefore, I believe we need to work to make them stronger and more just.

Political ecology

Political ecology emerged as a critical social science lens to better understand the drivers of the ecological crisis in the 1970s. While it may have different precedents and antecedents, my introduction to it has taken place within the fields of environmental studies and geography. With a nascent ecological crisis heralded by peak oil narratives, a cause for growing concern in the Western world about limits to growth, and the corresponding power and interest of the West in the resources and affairs of a newly postcolonial Third World, many environmental experts and technocrats were in the Third World trying to help solve 'land degradation' problems ranging from desertification in Africa to deforestation in South Asia (Robbins 2004). Problematically, many so-called environmental experts espoused neo-Malthusian assumptions about the reasons for these environmental changes – typically blaming the overbreeding tendencies of rural resource users in the Third World. Both the overbreeding and the overuse of natural resources were attributed to ignorance and selfishness (Blaikie 1985; Hardin 1968). Eco-managerialism aided by Western expertise was deemed the solution, corresponding to crisis narratives originating from a Western gaze. Eckholm (1975) exemplifies this in the context of Nepal, where his take on 'The Deterioration of Mountain Environments' published in *Science* contributed to the much-hyped Theory of Himalayan Degradation, which was later debunked, but not before making a mark on the country's environmental policies and legislation (Guthman 1997; Ives 1987). A number of factors enabled casual observations leading to a theory of ecological crisis where it was not warranted, such as extrapolating findings deduced from one geographical context indiscriminately to a broader geographical context or in other settings. Racial prejudice in how Third World people are seen in relation to their environments has also played a part in neo-Malthusianism and green Orientalism in shaping these discourses (Lohmann 1993; O'Connor 2016). Ehrlich's (1968) *Population Bomb* is a case in point.

In a climate where highly subjective and sometimes unfounded pronouncements about Third World environmental degradation could be made with impunity, we would find solutions that matched how the problems were depicted. Nepal used to have a Ministry of Population and Environment from 1995 to 2018. US-based organizations such as the Rockefeller Foundation, Malthusian ideology, and 'population advisors' played an influential role in shaping India's population policies in the decades after Independence (Connelly 2006; Minkler 1977). In such a policy climate, the much-needed intervention made by political ecology was to introduce and insist on a critical evaluation of the underlying structural drivers of environmental crisis that were being attended to, thus moving beyond surface manifestations of the problem to their root causes (Robbins 2004). Blaikie (1985), in discussing soil erosion as symptom, cause, and outcome of underdevelopment, sought to analyze both place-based and global political economic forces that impinged upon the resource practices of the maligned local land user. Blaikie and Brookfield (1987) went further to deduce a 'chain of explanations' that would situate land degradation within a landscape shaped by colonialism and capitalism.

The impetus of this newly formulated field of study, political ecology, then, was to explain environmental change in the Third World within a broad framework situated within a Marxist political economy analysis. In doing so, scholars increasingly theorized environmental change in the Third World as structurally created as a result of a conjoined history of colonialism and capitalism, the legacies of which continue to the present day in the form of neocolonialism and neoliberal capitalism in the Third World and settler colonialism and capitalism within the First World. Explaining why fishers or foresters or farmers operated in the ways they did due to the structural constraints of a globalized political economy of neoliberal capitalism in a world irreversibly transformed by colonialism – and where European ideas of Enlightenment would shape the way in which fundamental elements of the postcolonial 'development' project were conceptualized and operationalized; and where knowledge claims about environmental problems and their solutions could be made – became de riguer within political ecology (Robbins 2004).

Political ecological analyses of a wide range of environmental conflicts and natural resources, ranging from fossil fuels to forests, farmland, fisheries, and water, are all useful in unpacking the climate crisis – whether one wishes to understand the dynamics involved in mitigating climate change by reducing GHGs or by sequestering carbon in forests, or to understand the dynamics of vulnerability to climate impacts and adaptation capability. There is considerable overlap between research on the commons and political ecology, given the interests of both sets of scholars on natural resources, their use, governance, and conflicts over them. Contrasted with the seemingly apolitical orientation of research and writing on the commons, whose focus tended to be on institutions of governance and their structure, the focus of political ecology has always been on power. The question of who has power, access, and control over resources, who makes decisions and who is affected by them, whose ideas and knowledge of environmental change count are all key questions asked by political ecologists when they examine environmental change, conflicts, conservation, or discourse. The work therefore spans the gamut of social, political, economic, and cultural structures that shape and constrain the agency of human and non-humans in the environment: capitalism, colonialism, patriarchal structures, and, increasingly, racism (Keucheyan 2016). The initial overwhelming reliance on a structuralist class analysis would later make way for post-structuralist analyses engaging in postmodern approaches, postcolonial theory, and feminist theory not only limited to unravelling the materiality of the means and modes of production and reproduction but also unpacking the discursive (Robbins 2004).

Over the past decade or so of political ecological research on climate change, important work has been done on complicating the nature-society binary inherent in mainstream analyses of climate change that posits humans as acting on nature, altering the climate system, which, in turn, acts on human systems. The notion of climate change as a socio-natural problem created in conjunction with the many ways in which non-human nature is co-produced with human activity has been a particularly critical intervention on the tendency most climate adaptation policy-makers have to limit their work to finding ways in which to make socio-ecological systems less vulnerable to the depredations of a climatic system gone haywire due to anthropogenic interference. The notion of climate change as a socio-natural phenomenon does the dual work of specifying which social activities have been most influential in co-producing nature in its current avatar, accordingly referred to as capitalocene, plantationocene, manthropocene, and chthulucene (Haraway 2015; MacGregor 2019; Moore 2017), as well as problematizing the notion of enhancing resilience of communities to (an external) climate change acting on them. Rather, the social structure – a heteropatriarchal capitalism in a colonial context – is simultaneously understood to be both creating the climate crisis and creating or entrenching the social vulnerabilities of human and non-human systems to it (Sultana 2020; Taylor 2014). Not surprisingly, some of these insights overlap with an understanding of how social vulnerabilities are co-created when the commons are enclosed (Sovacool 2018). Political ecology work on climate change has therefore done valuable work to emphasize structural solutions to the climate crisis.

One serious gap in political ecological analyses on the climate crisis, though, is to contend with the problem at the scale of the global. It does so in some ways but not in others. The hallmark of political ecology – particularly the Third World variant, which is also the context in which most early work was accomplished – has been to situate the agency of the resource user within a structure shaped by global capitalism and colonialism. The impetus in the early days was to deflect blame from the rural peasant. In the contemporary context, in studies engaging with the climate crisis, the impetus is to problematize Band-Aid solutions in climate adaptation that focus on the agency of the individual or group vulnerable to climate change without addressing the structural drivers of vulnerability and of the problem of climate change. There have also been numerous studies critically examining carbon trading as a mitigation measure (e.g. Bumpus and Liverman 2010). These remain important critical interventions that focus on global forces impinging on the life chances and livelihood prospects in localized contexts. A still unfulfilled gap is to think of the power struggles between nation-states in the context of North-South conflicts over the atmospheric commons.

Geographers have reasonably taken umbrage with the categories Global North and South as being geographically inaccurate, but also for their state-centrism. These spatial and territorial imaginaries are indeed socially constructed, but that does not make them worthy of dismissal as figments of people's imaginations. Much like the understanding of the acknowledgment that material non-human nature exists, even as 'nature' is a social construct; and that racism is real and operates through insidious processes through which racial formation becomes institutionalized, even if race is a social construct, the core and the periphery of the global political economy have a materiality and associated discursive elements. There are power dynamics at play between representatives of the more than 200 nation-states in the world, as well as between blocs of countries with shared histories, even as countries and coalitions of countries are never monoliths. There is a privileging of local sites of action and change, and a corresponding obliviousness of or indifference towards international sites of negotiation, policy-making, and North-South politics in political ecological analysis of climate change that I want to push back against. Chapters Two to Four engage in more depth with this effort, drawing on a Gramscian analysis to bridge the gaps between traditional political ecology and global environmental politics. Questions that are critical to political ecology: –who has power, who decides, who benefits, and who loses in the struggle over the global atmospheric commons – can and should be asked for the North/South axis of difference, even as these are not static categories, and even if they are state-centric concepts vulnerable to the politics of scale. Meanwhile, more granularity is needed for understanding nature-society dynamics in contexts such as in Nepal that were not colonized by Europeans. Chapters Five and Six examine the forest commons in Nepal as they are incorporated into carbon trading by seeking to provide a nuanced analysis of the legacies of colonialism, caste, and patriarchy.

Inspired by feminist political ecologists (Rocheleau 2015; Sultana 2020) and postcolonial thinkers (Grosfoguel 2011; Spivak 2010), I have chosen to write in a critically reflexive manner that renders visible to the reader my positionalities – in terms of both social and epistemic locations – in the context of examining North-South climate politics as well as the implications of carbon trading for forest communities in Nepal. My analyses in this book are attempts to offer a theoretically and empirically informed, yet necessarily partial and situated knowledge of the complexities of seeking climate justice (Haraway 1998; Mansvelt and Berg 2010).

The dilemma of scale

Scale is an important but contested concept in geography. Marston, Jones, and Woodward (2005) argued that we should reject a priori categories of hierarchical scale in favor of a 'site ontology'. While unable to fully heed their call to "expurgate scale from the geographic vocabulary" (ibid, 420), I seriously take their invitation to privilege 'sites' over 'scales' to describe the spaces in which the governance of climate mitigation happens. My research has spanned sites such as the UNFCCC's COP-15 meeting in Copenhagen, labelled aspirationally as 'Hopenhagen'; the institutions of climate governance at the site of nation-state governance, such as ministries in India and Nepal; institutions of forest governance at the level of community forest user groups (CFUGs); households; workshops and meetings engaging representatives of institutions at these myriad sites; and interviews with a wide range of individuals spanning these various sites. These myriad sites correspond to significantly different configurations of actors, geographic locations, and resources that in my mind are usefully and appropriately characterized by the designation of the a priori categories of local (for CFUG), national (for ministries and departments, and country-specific NGOs), and global or international (for UN COP meetings and corresponding discourses about a global or international agreement pertaining to the atmospheric commons). The authors wonder whether the spatial imaginary of the 'global' correspond to any materialities. If we think about climate change as an issue pertaining to the global atmospheric commons, there is no doubt that there is a materiality that operates at a site or scale of the global. This is not to say that there exists a global atmosphere and a local atmosphere, but there does exist a global atmospheric commons, a global forest commons, and many local forest commons; and there is indeed a relationship between the global atmospheric commons and the local forest and other commons. Here, I find it unwieldy and unhelpful to use the term 'site' to denote the global atmospheric commons. However, I find it perfectly workable to use that term to denote the COPs at which the fate of the atmospheric commons is deliberated upon.

My invitation for political ecology to attend to climate negotiations at this international site of action is an invitation to recognize and acknowledge that power dynamics occur at this site, and the ways in which power is distributed at this site impinges on power dynamics at particular local sites. Marston, Jones, and Woodward (2005, 426) recognize that "there is a politics to scale, and whether we engage it or abandon it can have important repercussions for social action". The authors reasonably object to the hegemony and self-importance claimed by discourses speaking for or about globalization, global capitalism, global political economy, and the like – thereby relinquishing to the margins of obscurity critical processes that happen in particularistic contexts and sites. While taking this critique seriously, I find it necessary to refer to the scales of community, local, national, regional, and global even as these are socially constructed, to correspond to resource/environmental or identity configurations such as race, Indigeneity, class, and gender. To the extent that these social constructions are reified by institutions corresponding with these geographic imaginaries, they assume a materiality and discursive presence that I find impossible to work around. I will be deliberate in referring to 'sites' of action and discourse whenever possible, such as when referring to COP-15 as a site for inter-state deliberation, while referring to climate change as a problem that exists at global, regional, local, and other scales, with their corresponding resource commons and associated populations. In keeping with the notion of scale as socially constructed, it is helpful to think of how the representative trope that is scale becomes real to the extent that policies and political and economic relationships become institutionalized at particular sites.

A good example of this can be seen in how in the transition from Constitutional Monarchy to Federal Republic, Nepal went from having a centralized bureaucracy with five zones and 75 districts to a governance structure of seven provinces and 77 districts. Although the people and the landscape were the same, they were placed into new categories, and the newly constructed socio-political entities – particularly at the scale of province – became new arbiters of resource allocation and sites of decision-making, as well as a space for identity-based politics. Therefore even if the national, the provincial, and the local scales are not a given and may not truly exist, they exist as imaginaries and in the embodiment of institutions corresponding to these imaginaries, if only because the new Constitution of Nepal has willed them into being, replacing the older categories. Even if the newly devised categories do not accurately represent the people therein – and neither did the old – the institutions established to facilitate governance at each site constitute a meaningful space of policy-making, politics, and, by extension, scale of analysis. Notwithstanding the insight that nation-state identities are a territorial trap we often fall into (Agnew and Corbridge 1995), therefore, the site of the nation-state similarly constitutes a 'scale' of analysis as well as a politics of scale. Likewise, the institutionalization of states as sovereign entities and actors by the UN, as well as the UNFCCC as forum for the international deliberation over a fair burden-sharing agreement for the mitigation of climate change, renders actors with authority at this site – or the national scale – agency to make decisions and to negotiate with actors at other sites of action, decision-making, and discourse. The national scale mediates the resource flows and discourses between sub-national and international sites and scales, even as individuals and groups are able to 'jump scale' "to organize the production and reproduction of daily life and to resist oppression and exploitation at a higher scale – over a wider geographical field" (Smith 1992, 60).

In thinking of scale and the politics of scale in the context of climate change, two things merit attention. One is the question of justice. The other is the question of how the global atmospheric commons is related to myriad localized commons, and how to reconcile the two. The challenges posed by scale politics in navigating justice claims is the subject matter of Chapter Four. In Chapter Two I engage with the North-South framing of climate justice claims and their critiques. Rather than dwell on whether or not the scalar categories of Global North and South are valid, I examine the ways in which they are contested, negotiated, and defended, and the ways in which particular scale configurations are conjured up in counterarguments – a common one being that a class analysis of the global population is more accurate than the state-centric Global North and South, each of which contains outliers within, both in terms of nation-states as well as groups within. Chapter Three engages this discourse in a conversation with postcolonial theory. Chapter Four seeks to complicate the North-South justice argument by critically examining the 'right to development' claim embedded in a North-South climate justice argument, moving from international politics to the local context in Nepal. Chapter Five unpacks a carbon trading solution that implicates the forest commons in Nepal. In Chapter Six, I engage in a structural analysis of the forest commons and the ways in which they are becoming enmeshed in the commodification of the atmospheric commons. In Chapter Seven I suggest ways in which the global atmospheric commons is imbricated in myriad local commons and offer ways to conceptualize the taking back of the commons at multiple scales.

I argue for a both/and approach in lieu of an either/or approach to thinking about global justice: one that does not diminish and invalidate the existence of inequality and difference shaped by the legacy of European colonialism in shaping the current world order, even as the contours of new and emerging world orders and associated internal colonialisms are articulated. A multi-scalar analysis of commons and colonialisms enables us to tease out contradictions in particular justice claims. All categories are social constructions and have some material basis. The material conditions constitute the basis of climate justice claims that conjure up certain scalar or spatial configurations. Scale politics is therefore immensely central to the dilemmas of climate justice. There is a politics of scale and a politics of justice claims, and a need to understand how these politics intersect to create particular geographies of climate justice.

Figure 1.1 Copenhagen during COP-15, December 2009

Climate justice

Climate justice is an aspirational idea around which countries, non-profit organizations, grassroots movements, and youth worldwide have mobilized to draw attention to various forms of blatant injustice associated with climate change – how it disproportionately affects the most vulnerable communities around the world who have the fewest resources to withstand the consequences of a warming world while having contributed the least to causing it. Beyond this triple injustice (Roberts and Parks 2007), technological and market-based solutions designed to mitigate or adapt to climate change have also been known to further exacerbate vulnerabilities, leading to yet another dimension of climate injustice. Climate justice is an approach to address the climate crisis that is committed to minimizing and undoing existing forms of inequality and oppression and move towards a more socially just and ecologically sustainable world. This emergent idea has also led to an area of study dedicated to understanding the various meanings, movements, and policies associated with it. As a thematic area of study, climate justice is a multi-scalar and multifaceted concept that encompasses subjective notions about various forms of inequality that are conceptualized at different scales, whether in the creation of the climate crisis or in the context of mitigation or adaptation. Drawing on environmental justice as a field of study, arguments for climate justice often include corrective, distributive, and participatory justice approaches, and increasingly call for recognition and transformative structural change to address differences in power, privilege, and access to resources along lines of race, Indigeneity, class, and gender broadly construed (Walker 2012). As such, questions of responsibility for emissions, vulnerability to the impacts of climate change, representation in decision-making spaces both in international and local sites of action, and epistemological questions about how knowledge about climate change is produced and validated have been raised in the burgeoning climate justice literature. While structural analyses of climate change are often focused on the political economy of capitalism and colonialism, increasingly racism and heteropatriarchal structures are problematized as root causes.

During my decade-long period of learning about and teaching in this area in the United States, I have found that interest in the subject waxes and wanes with the prevalence of high-profile events and that the way in which it is represented in popular discourse tends to be shaped by the events. For instance, in the lead-up to the Copenhagen COP-15, the notion of a fair, ambitious, and binding (FAB) international treaty was prevalent, along with notions of climate debt seen as owed from Global North to Global South. Following the Arab Spring and the consequent popularity of the Occupy Movement and the notion of the 1 percent and the 99 percent becoming part of popular discourse, the 2014 People's Climate March in New York City congealed the idea of climate justice as synonymous with the disruption of the socially unjust and ecologically unsustainable system of capitalism. The Water is Life noDAPL (Dakota Access Pipeline) movement in 2016 was a moment when climate justice became characterized as an Indigenous-led movement for sovereignty and self-determination over resources vital to life as well as sacred sites, even as some leaders of the movement objected to calling it a climate justice movement. In 2020 with the salience of BLM, climate justice is increasingly being equated with racial justice. A gender-focused retelling of the struggle for climate justice has not achieved a comparable degree of prominence yet, although those ideas were put forth during the peak of the MeToo Movement (Johnson 2018) and have emerged in the recognition of women's reproductive labor during the pandemic (Power 2020). Likewise, although climate justice has not been explicitly characterized via a generational lens, except in a more general sense with the Brundtland Commission's definition of sustainable development, Greta Thunberg's highly visible and inspiring leadership – along with numerous Indigenous and BIPOC youth leaders worldwide – implicitly suggests there is a generational divide in responsibility, vulnerability, and capability.

The idea that justice and equity considerations are central to addressing the climate crisis has been central to Global South climate discourse decades now. Gradually a paradigm shift has occurred within the Global North, with the notion of climate justice becoming more prominent within climate movements in the United States, in my view. A justice framing is compelling because everyone has a subjective idea of what is just, but this is also precisely why there has been no consensus on a universal definition and why there have been many competing justice claims. The climate crisis has been known to be a 'super wicked' problem due to the sheer complexity of resolving the dilemmas posed by multiple and contradictory visions of what a just solution to the climate crisis is (Levin et al. 2012). When competing ideas of climate justice come to play, Schlosberg's (2009) insight about the importance of pluralism and bicameralism is useful to heed. In instances where different groups may have equally valid perspectives from their points of view and positionalities, the ability to be open to other vantage points is critical. Fostering a politics of avowal rather than a politics of disavowal helps build a culture of tolerance and respect for others' priorities and struggles (Kim 2015). The alternative perpetuates a culture of hubris which is part of the problem, not part of the solution. An emphasis on a singular vision of justice also perpetuates oversimplified and/or partial understandings of the climate crisis as opposed to a multi-faceted understanding of a complex problem. Walker's (2012) work on the politics of environmental justice claims – paying attention to a wide range of conceptualizations of environmental inequality, corresponding evidence offered to justify these claims, and theoretical explanations provided for why they exist – is one of the most helpful intellectual and pedagogical tools I have found for teaching students about how to think critically about climate justice claims. It is also an excellent entry point to starting to conceptualize the geographies of climate justice, since justice claims conjure up particular scalar and spatial imaginaries such that the politics of scale and the politics of climate justice are intertwined.

Within political ecology there has been an impetus to address structural drivers of climate change. This emphasis was present in critical work done in the field of environmental justice (EJ) in the 2000s that found the emphasis on questions of spatiality and distributive justice too limiting and argued instead for addressing structural impediments to justice, namely capitalism (Low and Gleeson 2002). This follows an earlier theoretical explanation of environmental racism for why communities of color and Native communities have been treated as sacrifice zones in the United States and globally (Mohai, Pellow, and Timmons Roberts 2009). Yet later EJ scholars would theorize racial capitalism (Pulido 2016), reproductive justice (Chiro 2008), and patriarchal structures as alternate explanations (Acha 2019; Buckingham and Kulcur 2009). While pointing to the need for structural transformations is warranted, it is also important to not invalidate demands for distributive justice made within the constraints of prevailing structures, as arguments for climate reparations are made by Global South countries to receive compensation for past harm and as a means to access financial resources for both mitigation and adaptation (Goeminne and Paredis 2010; Martinez-Alier 2002). North-South justice claims have often been problematized in favor of a class-based structural analysis of global climate injustice. I suggest three reasons such critiques are problematic: one, they do not reflect an understanding of abiding core-periphery power dynamics in international relations; secondly, when Global

North academics make these critiques, they come across as unreflexive and hypocritical given their own positionality and privileges within prevailing structures of carbon-intensive capitalist economies; and lastly, such either/or arguments are unnecessarily hierarchical in their approach to evaluating different justice claims, particularly in regards to disavowing Global South claims to ecological space.

Teaching climate justice in the US classroom

The first time I taught a class titled Climate Justice was in 2010 as a graduate student who was encouraged to propose a course not in the books (ENVS 411: Special topics). I have taught the subject three times since, at The Evergreen State College. One of the challenges I often face in teaching this program is how to move past issues such as climate denialism and climate anxiety, to try to unpack the multi-faceted, multi-scalar, and pluralistic concept of climate justice. Where we already have a triple injustice of responsibility, vulnerability, and capability (Roberts and Parks 2007); and the added dimension of unjust climate solutions that either exacerbate existing inequities and/or create new ones, the invisibilization of Global South experiences, priorities, and voices in US climate justice discourse – and the continued dominance of US experiences, priorities, and voices – constitutes yet another dimension of climate injustice. Likewise, de-centering the United States in conversations about climate justice in my classrooms has been challenging, in large part because students tend not to be able to relate or connect to the issues faced outside the United States or the Americas beyond a superficial – if polite – interest. I have come to sense that most students who care about climate justice tend to be overwhelmingly focused on issues of climate inequalities and injustice within the US context, whether focused on Indigenous populations or communities of color or low income communities. If there is an interest in the world outside the United States, it tends to be in the form of an exclusive focus on 'climate refugees', or of the impacts of US militarism or global capitalism, broadly construed.

With the momentum built by the reinvigorated BLM movement, calls for including a more accurate history and geography of slavery as well as that of settler colonialism have grown in the United States. In a similar manner, students in the country most responsible for climate change should also learn a more accurate history and geography of the climate crisis and the international process to address it. Teaching of climate justice must include not only the typical emphases on Indigeneity, race, class, and gender as important axes of difference, but also the North-South articulation, if only to pay homage to the historical context from which the climate justice discourse arose, but hopefully this would be a start to acknowledging the real existential crisis that climate change poses to communities in the Global South. In addition, understandings of Indigeneity, race, and gender in the Global North context often do not apply universally to the Global South; broader and more nuanced understandings are needed.

I am concerned that if we cannot move our concerns and our understanding of climate justice away from a place of US-centrism, then we are perpetuating the problem of US exceptionalism that has gotten in the way of a FAB international climate treaty. If we cannot conceptualize solutions to the climate crisis emerging from the Global South, and can only wrap our heads around climate solutions – whether in the form of eco-socialism, the Green New Deal, or a whole host of technological and market-based solutions – that originate in the United States or in the Western world or the Global North, we perpetuate the hubris of a Euro-centric world yet again – even as European colonialism and US imperialism in the form of neoliberal capitalism have compromised the capabilities for self-governance and self-determination of resource-dependent communities in the Global South in the first place. Nixon (2011, 2) refers to this as 'slow violence', by which he means: "a violence that occurs gradually and out of sight, a violence of delayed destruction that is dispersed across time and space, an attritional violence that is typically not viewed as violence at all".

It has been heartening to see the evolution of the nascent climate justice movement in the United States, but there is room for growth. Racial justice can be conceived not just as protecting racialized groups in the United States from acts of visible, explicit, and immediate violence, as well as the less visible forms of 'slow violence', compounded further by climate change. Just as we understand how institutionalized racism works in the United States, we need to understand how core-periphery relations institutionalized in global environmental and economic governance perpetuate the subjugation of racialized others in a global context, and need to be reformed to correct the structural imbalance that makes some countries more powerful than others, and more responsible for climate change. Just as we must problematize White supremacy when we talk about racism, we must also problematize US exceptionalism when talking about climate justice.

> What I call superpower parochialism has been shaped by the myth of American exceptionalism and by a long-standing indifference – in the US educational system and national media – to the foreign, especially foreign history, even when it is deeply enmeshed with U.S. interests.
>
> (Nixon 2011, 35).

Many in the United States have responded with outrage to police brutality towards Black and Indigenous communities in the US. I would argue that similar outrage is warranted for every person in the Global South who dies from climate change and other forms of violence perpetrated by the United States and other imperial/colonial nation. The United States has not acknowledged its historical responsibility for climate change and its debts to the Global South, and those involved in climate justice struggles in the United States should also not be oblivious to that question and how the United States should balance its obligations to marginalized communities within and without. This is a conversation that is long overdue. Environmental and climate justice movements and organizations increasingly recognize the importance of Indigenous knowledge, worldviews, and leadership in addressing the climate crisis, particularly in shaping how resilience is conceptualized and practiced. Here too, the global context is often marginalized in relation to the US context.

If we are to depart from the problematic notion that the United States and others in the West need to be saviors of the Global South that have been victimized by climate change and ironically by US hegemony and White saviorism, then learning from distant and racialized others is a must. Rather than end with a grand claim about neoliberal capitalism being the root cause of the climate crisis, and argue that eco-socialism is the answer, and the true form of it has never been practiced – and by implication it only exists in the minds of the theorists – I ask if we may consider existing vestiges of traditional commons management structures as models of the eco-socialism we want to see in the world, and then deliberate on what it might mean to scale up those structures, and/or what it might mean to connect them with other existing structures in ways that do not erode their strength or their essence. Even as the image of the Global South of the perpetual victim is problematic, it remains important to teach and learn about oppressive inequalities in the world system that exacerbate climate disparities. In much the same ways as we are witnessing a paradigm shift in how the world thinks about White supremacy today, we need a paradigm shift in how US exceptionalism and supremacy is viewed in the context of the climate crisis. An image of the United States as the knee on the neck of the Global South that has been subjected to US-led neoliberal capitalism, imperialism, militarism, and now by US-created climate consequences may be one we consider. And while the world outside the United States does not need this visual reminder, it may be one needed for those in the United States to react as powerfully as they have done to Floyd's death.

This book draws on research spanning the period of 2008–2019, including dissertation and post-dissertation work. The work is necessarily limited by my fieldwork in India, Nepal, and Copenhagen, as well as to my lived experience in the United States, but I believe the breadth I cover within these constraints helps shed light on the linkages between ideas of climate justice across multiple sites and scales of action and imagination, as well as the connections between the atmospheric commons and myriad grounded commons. I believe the discipline of geography offers invaluable insights in thinking of notions of space, place, and scale as they relate to the twin subjects of climate justice and the commons, and the contributions may be helpful to anyone interested in arriving at a nuanced understanding of their rich and complex intersections.

References

Acha, Majandra R. 2019. "Climate Justice Must Be Anti-Patriarchal, or It Will Not Be Systemic." In *Climate Futures*, edited by Kum-Kum Bhavnani, John Foran, Priya A. Kurian, and Debashish Munshi. London: Zed Books.

Agnew, John, and Stuart Corbridge. 1995. "The Territorial Trap." In *Mastering Space*, edited by John Agnew. London: Routledge.

Barnett, Jon. 2007. "The Geopolitics of Climate Change." *Geography Compass* 1, no. 6: 1361–75.

Berger, Mark T. 2004. "After the Third World? History, Destiny and the Fate of Third Worldism." *Third World Quarterly* 25, no. 1: 9–39.

Blaikie, Piers. 1985. *The Political Economy of Soil Erosion in Developing Countries*. London: Longman.

Blaikie, Piers, and Harold Brookfield. 1987. *Land Degradation and Society*. London: Methuen & Co. Ltd.

Buckingham, Susan, and Rakibe Kulcur. 2009. "Gendered Geographies of Environmental Justice." *Antipode* 41, no. 4: 659–83.

Bumpus, Adam, and Diana Liverman. 2010. "Carbon Colonialism? Offsets, Greenhouse Gas Reductions and Sustainable Development." In *Global Political Ecology*, edited by Richard Peet, Paul Robbins, and Michael Watts. London: Routledge.

Castro, João A. D. A. 1972. "Environment and Development: The Case of Developing Countries." In *Green Planet Blues, 2010*, edited by Ken Conca and Geoffrey D. Dabelko. Boulder: Westview Press.

Chiro, Giovanni Di. 2008. "Living Environmentalisms: Coalition Politics, Social Reproduction, and Environmental Justice." *Environmental Politics* 17, no. 2: 276–98.

Connelly, Matthew. 2006. "Population Control in India: Prologue to the Emergency Period." *Population & Development Review* 32, no. 4: 629–67.

Eckholm, Eric. 1975. "The Deterioration of Mountain Environments." *Science* 189, no. 4205: 764–70.

Ehrlich, Paul R. 1968. *The Population Bomb*. New York: Ballantine Books.

Ervine, Kate. 2018. *Carbon*. Cambridge: Polity Press.

Executive Order 13914 of April 6. 2020. "Encouraging International Support for the Recovery and Use of Space Resources." *Code of Federal Regulations* 20381–82, title 3.

Gibson-Graham, J. K., Jenny Cameron, and Stephen Healy. 2013. *Take Back the Economy*. Minneapolis: University of Minnesota Press.

Goeminne, Gert, and Erik Paredis. 2010. "The Concept of Ecological Debt: Some Steps Towards an Enriched Sustainability Paradigm." *Environment, Development, Sustainability* 12: 691–712.

Grosfoguel, Ramon. 2011. "Decolonizing Post-Colonial Studies and Paradigms of Political-Economy: Transmodernity, Decolonial Thinking, and Global Coloniality." *Transmodernity: Journal of Peripheral Cultural Production of the Luso-Hispanic World* 1, no. 1: 1–38.

Guthman, Julie. 1997. "Representing Crisis: The Theory of Himalayan Environmental Degradation and the Project of Development in Post-Rana Nepal." *Development and Change* 28: 45–69.

Haraway, Donna J. 1998. "Situated Knowledges: The Science Question in Feminism and the Privilege of Partial Perspective." *Feminist Studies* 14, no. 3: 575–99.

———. 2015. "Anthropocene, Capitalocene, Plantationocene, Chthulucene: Making Kin." *Environmental Humanities* 6: 159–65.

Hardin, Garrett. 1968. "The Tragedy of the Commons." *Science* 162: 1243–48.

———. 1994. "The Tragedy of the Unmanaged Commons." *Trends in Ecology & Evolution* 9, no. 5: 199.

Harvey, David. 2018. *Marx, Capital and the Madness of Economic Reason*. New York: Oxford University Press.

Islam, Arafatul. 2020. "Cyclone Amphan: Victim Feels 'Abandoned' by International Community." *DW News*, May 26.

Ives, Jack D. 1987. "The Theory of Himalayan Environmental Degradation: Its Validity and Application Challenged by Recent Research." *Mountain Research and Development* 7, no. 3: 189–99.

Johnson, Princess Daazhraii. 2018. "What's Missing from #MeToo and #TimesUp: One Indigenous Woman's Perspective." *Native Movement*, January 20.

Keucheyan, Razmig. 2016. *Nature Is a Battlefield*. Cambridge: Polity Press.

Kim, Claire Jean. 2015. *Dangerous Crossings*. New York: Cambridge University Press.

Levin, Kelly, Benjamin Cashore, Steven Bernstein, and Graeme Auld. 2012. "Overcoming the Tragedy of Super Wicked Problems." *Policy Sciences* 45: 123–52.

Lohmann, Larry. 1993. "Green Orientalism." *Ecologist* 23, no. 6: 202–4.

Low, Nicholas, and Brendan Gleeson. 2002. "Expanding the Discourse: Ecosocialization and Environmental Justice." In *Environmental Justice*, edited by John Byrne, Leigh Glover, and Cecilia Martinez. New Brunswick: Transaction Publishers.

MacGregor, Sherilyn. 2019. "Zooming in, Calling Out: (M)Anthropogenic Climate Change Through the Lens of Gender." In *Climate Futures*, edited by Kum-Kum Bhavnani, John Foran, Priya A. Kurian, and Debashish Munshi. London: Zed Books.

Mansvelt, Juliana, and Lawrence D. Berg. 2010. "Writing Qualitative Geographies, Constructing Meaningful Geographical Knowledges." In *Qualitative Research Methods in Human Geography*, edited by Iain Hay, 3rd ed. Oxford: Oxford University Press.

Marston, Sallie A., John P. Jones, and Keith Woodward. 2005. "Human Geography Without Scale." *Transactions of the Institute of British Geographers* 30, no. 4: 416–32.

Martinez-Alier, Joan. 2002. "Green Justice – Ecological Debt and Property Rights on Carbon Sinks and Reservoirs." *Capitalism Nature Socialism* 13, no. 1: 115–19.

Miller, Marian. 1995. *The Third World in Global Environmental Politics*. Buckingham: Open University Press.

Minkler, Meredith. 1977. "Consultants or Colleague: The Role of US Population Advisors in India." *Population & Development Review* 3, no. 4: 403–19.

Mohai, Paul, David Pellow, and J. Timmons Roberts. 2009. "Environmental Justice." *The Annual Review of Environment and Resources* 34: 405–30.

Moore, Jason W. 2017. "The Capitalocene, Part I: On the Nature and Origins of Our Ecological Crisis." *The Journal of Peasant Studies* 44, no. 3: 594–630.

Narain, Sunita. 2019. "Equity: The Final Frontier for an Effective Climate Change Agreement." In *Climate Futures*, edited by Kum-Kum Bhavnani, John Foran, Priya A. Kurian, and Debashish Munshi. London: Zed Books.

Nixon, Rob. 2011. *Slow Violence and the Environmentalism of the Poor*. Cambridge, MA: Harvard University Press.

O'Connor, M. R. 2016. "One of the Most Repeated Facts About Haiti Is a Lie: How Race Complicates the Way We View Haiti and the Environment." *Vice*, October 13.

Ostrom, Elinor. 1990. *Governing the Commons*. London: Cambridge University Press.

———. 1999. "Coping with Tragedies of the Commons." *Annual Review of Political Science* 2: 493–535.

Power, Kate. 2020. "The COVID-19 Pandemic Has Increased the Care Burden of Women and Families." *Sustainability: Science, Practice and Policy* 16, no. 1: 67–73.

Pulido, Laura. 2016. "Flint, Environmental Racism, and Racial Capitalism." *Capitalism Nature Socialism* 27, no. 3: 1–16.

Robbins, Paul. 2004. *Political Ecology*. Malden, MA: Blackwell.

Roberts, J. Timmons, and B. C. Parks. 2007. *A Climate of Injustice*. Cambridge, MA: The MIT Press.

Rocheleau, Dianne. 2015. "A Situated View of Feminist Political Ecology From My Networks, Roots and Territories." In *Practising Feminist Political Ecologies*, edited by Wendy Harcourt and Ingrid L. Nelson. London: Zed Books.

Schlosberg, David. 2009. *Defining Environmental Justice*. New York: Oxford University Press.

Skjaerseth, Jon B. 2012. "International Ozone Policies: Effective Environmental Cooperation." In *International Environmental Agreements*, edited by Steinar Andresen, Elin Lerum Boasson, and Geir Honneland. New York: Routledge.

Smith, Neil. 1992. "Contours of a Spatialized Politics: Homeless Vehicles and the Production of Geographical Scale." *Social Text* 33: 54–81.

Sovacool, Benjamin. 2018. "Bamboo Beating Bandits: Conflict, Inequality, and Vulnerability in the Political Ecology of Climate Change Adaptation in Bangladesh." *World Development* 102: 183–94.

Spivak, Gayatri C. 2010. "'Can the Subaltern Speak?' Revised Edition, from the 'History' Chapter of Critique of Postcolonial Reason." In *Can the Subaltern Speak? Reflections on the History of an Idea*, edited by R. C. Morris. New York: Columbia University Press.

Sultana, Farhana. 2020. "Political Ecology 1: From Margins to Center." *Progress in Human Geography* 1–10.

Taylor, Marcus. 2014. *The Political Ecology of Climate Change Adaptation*. New York: Routledge.

Vidal, John. 2010. "Confidential Document Reveals Obama's Hardline US Climate Talk Strategy." *The Guardian*, April 12.

Walker, Gordon. 2012. *Environmental Justice*. New York: Routledge.

Wettestad, Jorgen. 2012. "Reducing Long-Range Transport of Air Pollutants in Europe." In *International Environmental Agreements*, edited by Steinar Andresen, Elin Lerum Boasson, and Geir Honneland. New York: Routledge.

Williams, Marc. 2005. "The Third World and Global Environmental Negotiations: Interests, Institutions and Ideas." *Global Environmental Politics* 5, no. 3: 48–69.

2 North/South climate politics and the role of India

The seeds for this chapter – based on my PhD dissertation – were planted two decades ago when I was researching the question of whether or not Nepal should ratify the Kyoto Protocol, for a newspaper article I was writing for *The Himalayan Times* in 2001. This was my first full-time job fresh after graduating from college with a bachelors degree in environmental sciences, and my first foray into the world of global climate politics. Several months later, in 2002, I had an opportunity to attend a workshop for South Asian environmental journalists organized by the Centre for Science and Environment (CSE) in Delhi, where I was introduced to the subject of North/South environmental politics. Sunita Narain, director general of CSE, made a compelling case for the importance of equity in shaping global climate governance and opened my eyes to the skewed politics of climate change in international negotiations, where Global North countries fought to keep the equity agenda out of the negotiations. Several years would pass before I would return to the subject for academic research, spurred by a classroom experience. In 2007, as a third-year doctoral student and a graduate teaching fellow in the Environmental Studies Program at the University of Oregon, I had an opportunity to give a presentation on the politics of the Kyoto Protocol in an Introduction to Environmental Studies course for undergraduates. A lively discussion ensued. Many students – representing a sample of young people in an environmentally progressive state in the Pacific Northwest of the United States – questioned both the categorization of countries such as India and China as 'developing countries' and the unfairness of the Kyoto Protocol's provisions which mandated emissions reductions for the United States but not for India and China, sentiments that unsurprisingly mirrored the negotiating positions of the United States. In the process of facilitating this class discussion, I had found my dissertation topic.

I would examine climate negotiations from the point of view of a key nation-state advocating for a North/South framing of climate justice and seek to situate the discourse in the environmental justice and political ecology literatures in ways that were informed by postcolonial theory. I conducted in-person semi-structured key informant qualitative interviews in Delhi over a six-week period during November 2008–January 2009, participant observation and interviews over two weeks during COP-15 in Copenhagen, and analyzed NGO and government documents between 2008 and 2010. My primary sources for interviews were officials from the Ministry of Forests and Environment (MoEF); two influential NGOs in Indian climate discourse and policy, the Center for Science and Environment (CSE) and The Energy and Resources Institute (TERI), as well as other organizations, including Greenpeace India and the Indian Youth Climate Network (IYCN); and participants at COP-15. Officials interviewed included two delegates to UNFCCC negotiations and members of the Prime Minister's Council on Climate Change (PMCCC). As such, much of the analysis in this chapter is based on interviews with a small section of the Indian intelligentsia that was engaged in the discourse on domestic and international climate politics and deliberations in the build-up to COP-15. India's persistent emphasis on discourses of climate justice understood as North/South equity since 1990 (Agarwal, Narain, and Sharma 1999), coupled with its 'emerging economy' and 'major emitter' profile, makes this an important case study in global climate politics.

Like any modern nation-state India is not a monolith. The population of India is gendered, racialized, and classed, with different groups within the country positioned differently in terms of what is understood to be the triple injustice of climate change (Roberts and Parks 2007). India has its share of the wealthiest individuals with jet-setting lifestyles and high carbon footprints. It also has farmers, fishing and coastal communities, and urban slum dwellers that are severely affected by the impact of climate change on the water cycle and its consequent manifestations for rain-fed agriculture, unreliable fish catch, weather-related disasters, and heat waves. Its long disenfranchised *Adivasi* forest dwellers and other marginalized populations have engaged in decades-long struggles for rights and sovereignty over *Jal, Jangal, Zameen* – water, forests, and land – against an unjust state (Taylor 2014). Despite these intra-state disparities, India, a nation-state that emerged in 1947 after the end of Britain's colonial rule, is represented at UN climate negotiations as a singular political entity with a consistent negotiating position since 1990 and up to the present time. This negotiating position rests on its insistence that climate change is a global crisis precipitated by cumulative fossil fuel-burning activities of developed countries over the past two centuries, that India is particularly vulnerable to the impacts of climate change, and that historical responsibility and equity must be the twin pillars on which a fair global climate treaty must rest.

A North/South framing of climate justice in the Kyoto Protocol

The 'limits to growth' discourse that accompanied the 1972 UN Conference on the Human Environment has been a substantial source of North/South discord, where the Global North's environmental agenda was perceived by Third World environmentalists and intellectuals as a neocolonial strategy for the Global North to yield power over the Global South and to stunt its development (Castro 1972; Conca 2001; Meadows et al. 1972). Discussions of international EJ have consequently been framed in a North/South context, where the Global North was seen to owe an ecological debt to the Global South due to unfair appropriation of ecological space (Anand 2004; Martinez-Alier 2002; Srinivasan et al. 2008). The idea of ecological space originates from the 'limits to growth' discourse but departs from it by emphasizing equity in sharing limited space for growth. The North/South EJ argument is closely related to the North/South politics of the 1970s that arose from the context of proposals for a New International Economic Order (NIEO), where a coalition of formerly colonized countries sought to challenge the unfair terms of trade between themselves and industrialized countries, by demanding changes that would enable the Global South to achieve self-sustaining economic growth and industrialization (Bhagwati 1977; Najam 2004).

> The functioning of the UN during the last three decades and more has brought . . . the near abolition of colonialism. But economic domination of the old era has persisted. The economic patterns that exist continue to reflect the old controls. These we want changed. And the affluent and industrialized countries of the world do not want the status quo to be altered. Our effort has been to persuade the North. Some of them are idealistic and not totally negative. Others are not prepared even to think in terms of a consensus that can bind both the North and the South. It may appear that in the short-term we of the South have little leverage vis-à-vis the North. But in the long run, the meek but determined South cannot but gets its fair share of the good things produced by Mother Earth. We do not wish to supplant the present rule makers. We only wish to cooperate with them so as to ensure that the rules are not weighted against us for all time to come. We must take their legitimate interests and proper concerns into account. But they too must not pretend that we have no interests and cannot be permitted to express any concerns.
>
> (Indian bureaucrat, cited in Jha 1982, 72–73)

Gilman (2015, 5) asserts that "the NIEO was more than just a set of technical economic-legal proposals; it was also an explicitly political initiative, an attempt to extend the realignment of international power that the process of decolonization had begun". It was in this context that the 1992 UNFCCC and the subsequently formulated international climate agreement Kyoto Protocol emerged. The Kyoto Protocol was adopted in 1997 and came into force in 2005, with the mandate of requiring Annex I Parties to the Convention – consisting of 24 member countries of the Organization for Economic Cooperation and Development (OECD), the EU, and countries undergoing transition to a market economy (Agarwal, Narain, and Sharma 1999) – to reduce their GHG reductions by an aggregate of 5.2 percent of 1990 emission levels over the first commitment period of 2008–12. The reduction target was disaggregated for particular countries based on their emissions profiles. Flexibility mechanisms of emissions trading, Joint Implementation, and the Clean Development Mechanism could be used to meet these targets (Vogler 2016). Negotiations leading up to COP-15 had been framed as post-Kyoto negotiations in Western media, but after much negotiation and deliberation, the Kyoto Protocol was renewed for a second commitment period of 2013–20 during the Qatar COP-18 in 2012.

The texts of the UNFCCC and the Kyoto Protocol make two things clear: a clear demarcation between the responsibilities of developed and developing country parties, and the prioritization of economic growth and sustainable development for the latter. The UNFCCC, formulated during the 1992 Rio Earth Summit, established the parameters of differentiation "on the basis of equity and in accordance with their common but differentiated responsibilities and respective capabilities", adding that "the developed country parties should take the lead in combating climate change and the adverse effects thereof" (United Nations 1992, 4). The text of the convention explicitly notes that

> the largest share of historical and current global emissions of greenhouse gases has originated in developed countries, that per capita emissions in developing countries are still relatively low and that the share of global emissions originating in developing countries will grow to meet their social and development needs.
>
> (ibid, 1)

The Kyoto Protocol included explicit references to the responsibilities of industrialized countries to strive to meet their greenhouse reduction commitments to "minimize adverse social, environmental and economic impacts on developing country Parties" (United Nations 1998, 5). Other priorities that are stipulated are the establishment of funding mechanisms and transfer of technology, the continued emphasis on common but differentiated responsibilities (CBDR), and the emphasis on the pursuit of 'sustainable development' by all Parties. A clear departure from the UNFCCC was that it introduced binding commitments to reduce emissions for industrialized countries and European economies in transition, categorized in 'Annex I', and advised against the introduction of new commitments "for Parties not included in Annex I" (United Nations 1998, 9). This higher burden of mitigation for industrialized countries was based on the principles outlined in the UNFCCC text that sought to attribute historical responsibility for the current levels of anthropogenic GHG emissions to developed countries. Such articulation of differential responsibility is compatible with the North/South framing of (in)justice championed by scholars who speak of contraction and convergence (Agarwal, Narain, and Sharma 1999), ecological debt or ecologically unequal exchange (Martinez-Alier 2002; Srinivasan et al. 2008), and climate injustice (Roberts and Parks 2007).

Figure 2.1 Flyer outside the Bella Center, Copenhagen, December 2009

A brief overview of North/South politics in global climate negotiations from Rio to Kyoto and Beyond

Swedish scientist Svante Arrhenius documented patterns of a warming world due to rising carbon concentrations in the atmosphere in 1896, but it was not until the 1970s that the international community would be mobilized to pay attention. The first World Climate Conference took place in Geneva in 1979.

Scientists called for preventive action on climate change throughout the 1980s, but it was only after a series of natural disasters struck the developed world in the late 1980s – including a tropical storm in London in late 1987, the worst since 1703; and a drought in the United States in 1988, the worst since the 1930s – that momentum would begin towards formulating a global climate treaty in 1990, when the second World Climate Conference took place in Geneva, with 137 countries agreeing to negotiate a world climate treaty. The first such meeting took place in 1991 in Geneva, and the first agreement would be created in 1992 at the Rio Summit in the form of the UNFCCC, signed by 154 countries. The question of how to apportion responsibility for addressing the growing problem of climate change divided more and less developed countries from the start. The United States was a key obstructionist to a robust climate treaty with binding targets for stabilizing greenhouse gases from the start, resulting in US environmentalists denouncing the resulting Convention as a 'neutered' one (Agarwal, Narain, and Sharma 1999, 42).

In the lead-up to the Kyoto Climate Conference in December of 1997, the fossil fuel lobby in the United States issued full-page advertisements in newspapers across the United States – in October and November – decrying the impending Kyoto Protocol and urging then president Clinton to refrain from signing it (ibid). The United States would be the only country listed in Annex I to never have ratified the Kyoto Protocol. The 1997 Byrd-Hagel resolution – that passed the US Senate with a bipartisan consensus of 95–0 in June 1997 – prevented the United States from signing the Kyoto Protocol without similar commitments from India and China (Vogler 2016). Aside from obstructionism, the US position in climate negotiations has long been firmly against North/South equity, and insistent on comparable constraints for the major developing countries, particularly India and China, before it would accept responsibility. A major rationale for the United States' reluctance to ratify the Kyoto Protocol was to not recognize the validity of the Annex I/non-Annex I distinction for imparting mitigation responsibility (Desombre 2002; Najam, Huq, and Sokona 2003). Such reasoning resonates with the ideologies of 'race neutrality' and 'color-blindness', based on which claims of 'reverse racism' have been made to invalidate affirmative action policies in the US (Omi and Winant 2015) as it is a strategy to maintain dominance by invalidating a logic of differentiation along North/South lines whose aim is distributive justice and decolonization.

The Group of 77 (G-77) has long carried out the function of representing the interests of the Global South (Vihma 2010; Williams 2005), specifically the demand that the Global North is held accountable for its responsibility to the wider world for environmental degradation (Dodds 1998b). The G-77 was formed in 1964 as a coalition of 77 self-identified developing countries that has since grown to 134. Albeit with some points of internal discord, they have firmly demanded that Annex I nations take the lead in accepting aggressive emission reduction targets, based on historical responsibility as well as their own rights to development (Barnett 2007; Williams 2005). Other common South positions in global environmental negotiations are for the North/South transfer of financial resources and technological assistance for environmental programs above and beyond official development assistance, capacity building, environment-development linkages, and a longer time frame for implementation of new regulations for the South (Najam, Huq, and Sokona 2003; Williams 2005). These priorities have been institutionalized through global negotiations at various points in time.

The 1992 United Nations Conference on Environment and Development (UNCED) – also referred to as the Rio Earth summit – established explicit linkages between environment and development issues and where differential obligations of developing and developed countries were articulated. The international environmental treaties that followed have had consequently emphasized North/South dimensions of equity paired with a sentiment of corrective justice, as the 'polluter pays principle' denotes (DeSombre 2002, 10). Even before UNCED, the 1987 Protocol on Substances that Deplete the Ozone Layer (Montreal Protocol) created a precedent for differential treatment towards developing countries in the context of negotiations over the atmospheric commons (Rajan 1997). It instituted a financial transfer mechanism, the Multilateral Fund that developed countries contributed to, and a time lag to large and rapidly industrializing developing countries such as China and India. The Basel Convention on the Control of Transboundary Movements of Hazardous Wastes and their Disposal also created a space for a North/South frame for debates and negotiations where Third World states articulated a common position despite their heterogeneity in levels of development (Miller 1995). Similar North/South environmental politics have manifested in the context of climate change negotiations.

Efforts to bring the larger developing countries such India and China have been consistently made by negotiators representing Annex I countries in the lead-up to the UNFCCC, and again during post-Kyoto negotiations. During 2008 COP-14 in Poznan, several Annex I countries submitted proposals for differentiation within the non-Annex I group, which were also resisted by India and China (Vogler 2016). Anticipation of the 15th COP was significant, with expectations that the annual UN meet that year would produce the successor to the Kyoto Protocol, although my respondents painstakingly pointed out to me that from their point of view, the negotiations at COP-15 were meant to determine mitigation targets for Annex I countries for the second commitment period after 2012. Yet, in Copenhagen, there were intense debates over the viability of the Kyoto Protocol over issues of differentiation, and eventually the infamous Copenhagen accord was produced, which did away with the CBDRRC as well as the Annex I/non-Annex I distinctions of the Kyoto Protocol. This was not a legally binding document and has been highly critiqued for being a step backwards. Amid discontent within Indian civil society, India 'took note of' the Copenhagen Accord.

One of the points of contention during COP-15 was the issue of monitoring, reporting, and verification (MRV). Annex I negotiators argued that MRV should accompany any aid given for adaptation and mitigation pursuits, but this was strongly resisted by India and China, among other countries. Eventually in an unpopular (in India) move by India's Environment Minister Jai Ram Ramesh, India agreed to MRV for funded projects. The minister's role in the global climate negotiations was seen to be increasingly in the direction of internationalization, seen by some as a positive move away from its traditional position (Vihma 2011). Gupta, Kohli, and Ahluwalia (2015) note that after Copenhagen, India's stance on formerly non-negotiable matters on equity have softened and that it has even accepted voluntary mitigation targets, although these have not been absolute emissions caps, but rather GHG intensity targets, which are based on a measure of GHG reduction per unit of economic growth, rather than a measure of absolute reductions, therefore are weak from the point of view of mitigation needs and strategic from the point of view of protecting economic interests. Even as they were recalcitrant about accepting mandatory emissions caps, India, China, and the United States pledged different degrees of GHG intensity targets. In the build-up to COP-15, India made a commitment to reducing its GHG intensity by 20–25 percent from 2005 levels by 2020. After the Paris Agreement, it has pledged to reduce GHG intensity by 33–35 percent of 2005 levels by 2030. Another key development in Copenhagen was the formation of the BASIC group, comprising of Brazil, South Africa, India and China, whose major arguments were to emphasize the importance of the Kyoto Protocol, as well as to claim their right – as developing countries – to pursue sustainable development.

Meanwhile, the *Guardian* reported on the basis of leaked internal correspondence of the Obama administration that the US negotiating team adopted a deliberate strategy of differentiating between two kinds of developing countries and that "advanced developing countries must be part of any meaningful solution to climate change including taking responsibilities under a legally binding treaty" (Vidal 2010). The United States has over time been successful in garnering increased support from other industrialized countries in advancing this argument that goes back to the late 1980s when climate negotiations began. Annex I Parties such as the European Union had initially urged stronger mitigation commitments and were sympathetic to Global South perspectives on the need for differentiation but would eventually buckle under pressure to appease the United States (Agarwal, Narain, and Sharma 1999). The differentiation embedded in the Kyoto Protocol as an equity principle would remain absent in the articulation of the 2015 Paris Agreement, which became official in November 2016. According to this successor of the Kyoto Protocol, no country has a mandated limit on emissions, but rather, each signatory country submits a voluntary Intended Nationally Determined Contribution (INDC) statement, which should be updated every five years with progressively ratcheted up ambitions. While the CBDRRC is mentioned in the document as a guiding principle for determining commitments, it does not specify what is meant by it and how it is to be operationalized. Developed and developing countries are signified as having different roles to play in addressing climate mitigation and adaptation efforts – the mandate of developed countries is to 'lead' and to 'support' developing countries, but no explicit equity imperatives or directives are included. Further, it differentiates developing countries into categories such as most vulnerable and least developed such that the viability of the non-Annex I group is further negated (United Nations 2015).

India's role in advancing a North/South climate justice Agenda

India's championing of a North/South framing of climate (in)justice continues into the era of the Paris Agreement even as the clear formulation of equity principles no longer exists:

> India's INDC is based on the 1992 convention. . . . Our objective is to establish an effective, cooperative and equitable global architecture based on climate justice and the principles of Equity and Common but Differentiated Responsibilities and Respective Capabilities, under the UNFCCC.
> (n.d., p. 3)

India's advocacy for equity along North/South lines has stayed consistent since the early stages of climate negotiations when representatives of the Indian government lobbied for the development of a common position for the South as the G-77 representative during the landmark UNCED in 1992. This position included the following stipulations: global environmental concerns should not impinge upon programs for economic development or influence trade policies, and developed countries should take primary responsibility for corrective action in stemming global environmental pollution. This core position in international climate negotiations reflects a long-standing emphasis of developing countries that the responsibility for addressing contemporary global environmental problems falls on the North. It was further enabled by the institutionalization of the CBDRRC norm on the base of North/South disparities in the climate agreement (Williams 2005). Indian negotiators were instrumental in crafting the principle of CBDRRC and the insertion of statements in the text that attributed historic emissions of GHGs to developed countries (Jakobsen 1998).

Other efforts that were not incorporated in the framework convention were proposals for a convergence of country emission levels around a common per capita level, explicit articulation of the polluter pays principle (Agarwal, Narain, and Sharma 1999; Paterson 1996), and ensuring that the UNFCCC's funding mechanism would not be dominated by a North-dominated institution (Jakobsen 1998). Indian participation in a range of global environmental negotiations in the 1990s was characterized by a consistent position focused on advancing its interests regarding state sovereignty, exercising more power in international affairs, and North/South economic equity, as Global South interests, often assuming a position of leadership within the Third World coalition (Rajan 1997). Even with the alleged bargaining power afforded by the nature of global environmental problems to the Global South, the ability of Indian and other Southern negotiators to define the agenda of the negotiations has been thought to be compromised due to structural inequalities (Jakobsen 1998; Rajan 1997). Climate negotiations have been seen as a possible springboard for a reinvigorated NIEO-esque North/South agenda to revive the impetus for structural transformations in core-periphery relations (Gilman 2015; Joshi 2013; Najam 2004).

The climate discourse in India was strongly influenced by two NGOs, CSE and TERI, since the start of negotiations, with stronger positions on fair burden-sharing than of other Global South countries (Agarwal, Narain, and Sharma 1999). The formation of the PMCCC in 2007 is understood to have introduced more flexibility in its negotiating stance (Vihma 2011), while showing receding influence from CSE in developing climate policy. Although opinions among Indian climate policy intelligentsia have diverged (Dubash 2009), the core negotiating positions of the official Indian position on climate change, of refusing an absolute cap on emissions or a peak year, and demands for financial and technological transfer have been stable up to the present moment. The official Indian position in climate negotiations in the context of burden-sharing for mitigation has long emphasized the historical responsibility of Annex I states (Jakobsen 1998; Joshi 2013; Rajan 1997). India's 2008 National Action Plan on Climate Change prepared by the PMCCC attributed responsibility for climate change to the "accumulated greenhouse gas emissions in the atmosphere, anthropogenically generated through long-term and intensive industrial growth and high consumptive lifestyles in developed countries" (GoI 2008, 1). The document asserts that the "overriding priority" of developing countries – including India – is to pursue "economic and social development and poverty eradication" (ibid). Indian negotiators played a significant role in the creation of a Bali Action Plan (BAP) during the 2007 COP-13 negotiations held in Indonesia. The BAP stipulated that non-Annex I countries would pursue non-binding 'nationally appropriate action plans' (NAPA) to mitigate climate change, with financial and technological support from the Annex I countries. In India's INDC submitted in response to the 2015 Paris Agreement, these sentiments remain. The document reiterates India's commitment to equity and CBDRRC, emphasizes the stark difference in per capita emissions as well as resource gaps between developed countries and India, and indicates a desire and need for financial and technological resources from developed countries to enable it to meet its obligations of poverty eradication and low carbon development aspirations.

Interviews with Indian climate negotiators and officials

Indian officials and negotiators have consistently resisted the increasingly frequent discussions of current and future emissions of emerging economies and discourses of differentiation within non-Annex I (Vihma 2011). In light of arguments for differentiation, Indian statespersons and civil society representatives have vociferously defended the legitimacy of the distinction of the Global

South and North, frequently summoning the CBDRRC principle as stated in the UNFCCC text and emphasizing the importance of considerations of equity, in making their arguments. I have argued that India's official position on climate justice at the international level has been heavily influenced by the stronghold of North/South imaginaries and that self-identification with the Global South is a key element of India's climate politics (Joshi 2013). During interviews, Indian officials were keen on portraying India as a developing country with immense poverty, energy access challenges, and acute vulnerability to climate change. Western portrayals of India's emerging economy – and emerging emitter – status were seen as oblivious to these 'ground realities', and the proposals for differentiation between emerging economies and others within the non-Annex I group of countries were seen as a 'divide-and-conquer' strategy often attributed to British colonial rule (Bose and Jalal 2011). Indian intelligentsia also shared a keen sense of the climate crisis as an opportunity for the unequal status quo of the world to shift in favor of the Global South – in favor of a multi-polar world order – that this was threatening to countries in the Global North and that they had not acted in good faith in the climate negotiations:

> We feel that the Annex I Parties are not pursuing an honest environmental agenda. We believe that they are pursuing an economic agenda, and a strategic agenda, in the guise of an environmental agenda. . . .
>
> We consider that the fundamental environmental justice is equal per capita rights of global environmental resources after taking into account the fact of historic asymmetries in the use of these global environmental resources. . . .
>
> While the problem has manifestly been caused by developed countries, and it is their overuse of fossil fuels resources, that are aggravating this problem, we consider that it is farfetched and certainly not at all conducive to environmental justice to ask developing countries to curb their growth to address a problem which they have not created, and which they're suffering from.
>
> (Interview, Dr. Prodipto Ghosh, Distinguished Fellow at TERI, 2001–12 member of India's negotiating team at UNFCCC, December 2008, Delhi)

Beyond seeking Global North accountability for their cumulative encroachment on the atmospheric commons, the Global South position has also been focused on using the greater bargaining power enabled by the climate crisis to call for international distributive justice in an environment of systemic power imbalances between the core and periphery of the global political economy. As such, climate negotiations are seen as a forum for a reinvigorated discussion about the need for a new international economic order (Gilman 2015). My interviews reflected a similar sense of marginalization of states – including India's – in global economic and environmental affairs. It is noteworthy that

Figure 2.2 Dr. Prodipto Ghosh, Delhi, December 2008

conversations revealed that in the minds of the Indian intelligentsia, India represented the Global South, and the United States represented the Global North, with constant comparisons made between the two. In this imaginary, India's placement in the Global South is legitimate due to the absence of basic necessities in the lives of the majority of its population. It was incredulous to most people that the two countries should be seen as rivals or competitors bound to the same set of global obligations regarding climate mitigation. In the words of one journalist interviewed:

> You either have a concept of every citizen is equal or, they're not. Or there is some amount of difference that you agree. Then you need to admit to that, that an average American or an average US citizen is at a different level than an average Indian citizen. . . . But, the entire convention is based on the fact that every citizen is equal. Everyone's right should be protected by each and every government, internationally and nationally.
>
> (Interview, male journalist, December 2008, Delhi)

The argument for transfer of technology to enable India to pursue a low–carbon path to development was a significant issue for most people interviewed.

I think a good argument would be, that a give and take, means we take some responsibilities as India, and, we are facilitated in some ways because so many new technologies have evolved in western developed countries, they could maybe subsidize these technologies and facilitate us being able to handle our manufacturing sectors on consumption sectors, in energy, much more efficiently, much more in a clean manner, so it has to be transfer of technologies, fundamentally from the West, and India taking several commitments to be able to contribute to, as you said, mitigating this whole threat of global climate change, driven primarily by pollution, and production of industrial sector, and energy consumption. So it has to be bilateral between India and developed world.

(Interview, male TERI official, December 2008, Delhi)

A professor at Jawaharlal Nehru University (JNU) argued that more advanced greener technologies, currently owned by Northern countries, should be "shared as a common treasure, common heritage" or "common human heritage technology" apposite to the severity of the climate crisis (Interview, December 2008, Delhi). Others suggested a revision of the Intellectual Property Rights regime to explore the possibility of making such technologies into 'public goods' – a global technological commons parallel to the global atmospheric commons. These sentiments are reflected in India's INDC, which also references Gandhi, the Vedas, and the Indian Constitution to claim a predilection for environmentally conscious development:

If climate change is a calamity that mankind must adapt to while taking mitigation action withal [sic], it should not be used as a commercial opportunity. It is time that a mechanism is set up which will turn technology and innovation into an effective instrument for global public good, not just private returns.

(n.d., 2)

The notions of historical responsibility and per capita equity that framed India's negotiating positions on climate from the start were very present in my conversations with Indian officials, as reflected in these statements from a CSE official I interviewed in Delhi, and again in Copenhagen the following year:

I see role of India as a proactive nation, which is willing to engage proactively in the climate debate, and, which understands its capability, capacity as well as its responsibility. Now when I use these three terms, you have to understand that climate is not a simple environmental problem; it's an ethical issue, it has equity involved. And it is a climate justice issue. People who have created this problem are not going to suffer from this problem, as much as people who have not. So poor are likely to suffer, developing countries are likely to suffer more from the developed ones. And climate change is not the problem created by developing countries. First thing that you need to understand very clearly. . . .

Therefore when I say it is an issue of environmental justice, I essentially mean that the principle of polluters pay principle, must be central to the whole debate of climate change. We cannot say today, that you know, what you have done in the past, forget about it and let's move ahead. That can be one stand, but I think it is a very inequitable stand. Because developed world has become rich based on carbon based economy. They've used far more environmental space than what they're entitled to. So now it is the time they should reduce their emission. . . . There has to be dematerialization in the West. So I think climate debate is very important, the concept of contraction and convergence is very important. It basically tells you that everyone has minimum entitlement. If you have exceeded that entitlement then let's come down, and those who are below it can go up.

(Interview, December 2008, Delhi)

In Copenhagen, the grounds for the erasure of the Kyoto Protocol were laid, including a plenary session where a proposal to amend its Annex I/non-Annex I binary distinction was debated. India and China came under significant pressure as emerging economies and emitters to accept more ambitious climate mitigation targets. In my conversations with Indian officials, it was clear that the insinuation that India's emphasis on historical responsibility of the United States was taken to mean that India does not worry about the future or about its own responsibilities:

Of course when India talks about historical responsibility, nobody is saying forget the future. We are just saying don't forget the past.

(Interview, Indian scholar, male, December 2009, Copenhagen)

As far as historical emissions go, at no point in time is any of these countries saying that they will not mitigate. What they are saying is, you accept and take on the responsibility, of what you have done, and we will take on the responsibility to make sure that we will not go your way in the future. So at no point in time has India been heard saying that we will go the American way and we will do exactly what America or Europe has done. Everybody realizes that that can't be done. That argument flies in the face of what actually the countries are saying.

(Interview, Indian INGO official, female, December 2009, Copenhagen)

Given how important the notion of historical responsibility is to India and many others in the Global South, it is remarkable that Global North countries – most of all the United States – have been resistant to acknowledge such responsibility and accepting accountability for the creation of the problem. Agarwal, Narain, and Sharma (1999) noted that the United States fought strenuously since the early stages of climate negotiations to circumvent language in the Convention indicating historical responsibility for contributing to climate change, instead choosing to interpret CBDRRC with an emphasis on developed countries' capacity to lead and to offer resources. The stark difference across Global North and South perspectives on the question of historical responsibility suggests a form of cultural parallax at play, a phenomenon described by Nabhan (1998, 266) as "the difference in views between those who are actively participating in the dynamics of the habitats within their home range and those who view those habitats as 'landscapes' from the outside".

Fissures and unity within the G-77 fabric

During a plenary session at the COP-15 in the Bella Center in Copenhagen, Tuvalu proposed an amendment to the Kyoto Protocol's binary classification of countries – reflecting demands for tougher targets for the emerging economies by small island states – while emphasizing that its intent was not to replace the Kyoto Protocol but rather to strengthen it. This proposal saw a curious bifurcation among the G-77 and China negotiating bloc, with India, China, South Africa, and many other non-Annex I countries opposing the proposal while others supported it. Then MoEF Minister Jai Ram Ramesh emphasized the importance of coordinating a common negotiating positions and presenting a united face and added that India supported the G-77 position of upholding the UNFCCC, the Kyoto Protocol, and the Bali Action Plan. An Indian NGO official, female, pointed out that Tuvalu (not a G-77 member) was being represented by an Australian lawyer, and even if their intentions may have been good, the proposal had the effect of "unintentionally taking the pressure off of the developed world and putting the pressure on the developing world" (Interview, December 2009, Copenhagen).

In the face of increasing discussions about differentiation within the non-Annex I group and questions about the validity of the Global South as a meaningful category or the G-77 and China as a negotiating bloc, many interviewed were defensive:

> Why the differentiation first of all? You see, there are, two categories of countries: Annex I, and non-Annex I, which in the convention itself, we had agreed in 1990 that there are only two groups. Those who are listed as countries in Annex I, and those who are not listed. So either you are a developed country or you are a developing country. There is no need to create any further categories of countries, and the United Nations system does not recognize it.
>
> (Interview with MoEF official, male, December 2008, Delhi)

An awareness of the heterogeneity within these categories and the inevitability of recalibrating groupings based on level of development was certainly present.

But the belief that even in a system where the non-Annex I group was differentiated, India should continue to be regarded as a developing country for the purpose of climate negotiations was strong. Several who were interviewed suggested that the status of India as an emerging economy was hyped by Western media, since India was seen as a threat, in part because of its size. Yet others pointed out that whether India was a developing country or not was not the point; adhering to an agreement formulated after intense negotiations was more important; and that the graduation of countries along the development continuum does not alter historically contingent arrangements. Responding to questions about India's emerging economy status, a key negotiator clarified that the differential responsibilities arise from a specific context rather than perceptions of the level of development of countries: "the Annex I/non-Annex I distinction has nothing to do with developed/developing, but with level of CO_2 emissions in the 1990s" (Interview, Dr. Ghosh, TERI, Delhi, December 2008). An academic and environmental activist, male, affiliated with Jawaharlal Nehru University, emphasized the importance of understanding the systemic nature of North/South differences:

> The North has much higher average standards of living. . . . And in terms of the wealthy elite, India has the largest number of billionaires – someone has written – in the world. But if you look at the entire corpus of wealth, the top five percent of India, it's much less than the corpus of wealth the top five percent in the States or Britain or France or Germany, have. So there is no comparison. The point is well taken that everyone in the South is not poor. That's true. But it is also a system, where large parts of the South, the poor are getting poorer. That's not the case in the States, or in Northern countries. So there are elites, and there is poverty, but the kind of mass poverty [is not there].
>
> (Interview, December 2008, Delhi)

Seemingly justifying calls for differentiation of G-77, the dynamics of COP-15 in Copenhagen revealed an array of coalitions, such as BASIC (Brazil, South Africa, India, and China), the Alliance of Small Island States (AOSIS), the African Group, the Bolivarian Alliance for the Peoples of Our America (ALBA), and Least Developed Countries (LDCs). Such fissures led some to question the validity of the G-77 and China as a negotiating bloc and to claim that "(t)he geopolitics of the UNFCCC cannot be explained merely as a matter of the differences between the 'North' and the 'South'" (Barnett 2007, 1367). Vihma (2010, 3) argued that the raison d'être for the G-77 had ceased to exist given the "'elephant in the room' – an awkward balance between the political expediences of alliance and the prevalence of multiple voices of self-interest within the G-77 in the climate debate". But some of these coalitions have existed since the early 1990s. AOSIS, for instance, was formed in 1990, and from the earliest stages of global climate negotiations, it had pushed for a more ambitious treaty with stronger mitigation policies from the start (Agarwal, Narain, and

Sharma 1999; Rajan 1997; Vogler 2016). BASIC was formed in preparation for COP-15, but with the objective of pushing to save the Kyoto Protocol and its CBDRRC principle. Thus while the G-77 is not a homogeneous group and has shown fissures over time, with coalitions each with their own agendas, they have continued to support the core G-77 position and have often stuck together as a formidable negotiating bloc. This is especially the case for negotiations over the atmospheric commons such as with the ozone and the climate negotiations. The Like-Minded Group of Developing Countries (LMDC), which was formed in 2012 to coordinate positions among G-77 members from different coalitions listed earlier, was highly focused on preserving the sanctity of the non-Annex I grouping (Vogler 2016).

Although the G-77 has always been and remains heterogeneous, the logic of this coalition derives from the strength in numbers it provides to traditionally marginalized countries (Najam, Huq, and Sokona 2003), particularly in the context of UNFCCC negotiations, where the veto system gives de facto power to powerful Parties (Barnett 2007). India, China, and Brazil – states that have been most actively engaged in forging a common G-77 position – are ironically increasingly viewed as emerging economies and major emitters. In climate negotiations, the G-77 has remained consistent over time in sharing common priorities: upholding CBDRRC as an equity principle, pursuit of

Figure 2.3 Plenary session on proposed Kyoto Protocol amendment, Bella Center, December 2009

sustainable development, increased funding (in addition to overseas development assistance), transfer of technology, and enhancing the adaptive capacities and resilience of communities and states, particularly the most vulnerable ones (Najam, Huq, and Sokona 2003; Vogler 2016; Williams 2005). A persistent atmosphere of distrust between North and South Parties has also permeated climate negotiations (Najam, Huq, and Sokona 2003; Rajan 1997; Roberts and Parks 2007) precipitated in part by the North's failure to meet past commitments and obligations (DeSombre 2002). A strong conviction that climate change is a problem that has been precipitated by the economic growth process of the "early industrializers" who in turn wish to constrain the economic growth aspirations of the developing world has united an otherwise eclectic grouping of countries (Agarwal, Narain, and Sharma 1999; Williams 2005). These interests and convictions continue into the Paris Agreement, as India's INDC reveals.

Unpacking the Global South imaginary

The dissolution of the Annex I/non-Annex I divide is very much tied to the negation of the idea of the Global South as a coherent and valid political identity. The Global South – often used interchangeably with the terms 'Third World' or 'developing countries', and in contradistinction with the Global North or developed countries – often refers to countries that have been exploited by colonial powers (Anand 2004; Isbister 2006). As such, they are marked by a core–periphery relationship as a result of the legacies of colonialism, even after successful anti-colonial movements. Newly independent countries from Asia and Africa formed the Third World Project at the 1955 Bandung Conference and the Non-Aligned Movement in 1961 to indicate their preference for an independent way forward in the midst of the Cold War rivalries between the Soviet Union and the United States, and their respective allies (Isbister 2006; Prashad 2014; Williams 2005). The G-77 was established in 1964 to represent them at the UN. In their efforts to extricate themselves from the vestiges of a colonial world order, they put forth the NIEO proposal for North/South justice in 1974 (Gilman 2015; Jha 1982). Although the NIEO would not come to fruition, the idea of the South has been congealed as the embodiment of those "who are disenfranchised in an international system dominated by the industrialized countries: the North, the developed, the rich" (Isbister 2006, 16). A number of critical geographers and other scholars have critiqued the validity of these categories for their inability to represent global inequality accurately (Sowers 2007; Barnett 2007; Toal 1994a, 1994b).

When I use the term 'Global South' or 'Global North', I certainly am not referring to countries that are below or above the equator, although there is considerable overlap. As Anand (2004) has clarified, these terms when used in a North/South politics context connote political categories – shaped by the exigencies of a colonial and imperial history – rather than a physical geographical one. Further, she suggests that while each category is internally heterogeneous, their similarities outweigh the differences, and that in the context of global environmental policy and politics, the categories Global North and South continue to hold meaning. Isbister (2006, 16) sees these categories as representing the "opposition between the poor and the rich" as well as "the promise that those who are currently oppressed will eventually overcome their oppression". I suggest that we consider these categories not as pre-given and fixed containers of the world but rather as socially constructed categories that are contested and reinscribed for particular ends. Consequently, I see the Global South as a spatial imaginary that represents marginalized worlds (Joshi 2014). This includes countries that have been marginalized in the global political economy and those who are marginalized within the Global North (Escobar 2004; Williams 1993).

The North/South question in critical geo-politics

A North/South lens for representing global inequality has been challenged for various reasons, based on problematizations of the Global South imaginary (McFarlane 2006). For some scholars this has to do with its association with the Third World Project, no longer seen as meaningful or relevant after Third World states abandoned their radical positions to be co-opted by capitalist states, to the point of being "intellectually and conceptually bankrupt" (Berger 2004, 31; Slater 1997). Economic globalization had increased economic differentiation and fragmentation of the Global South, challenging the legitimacy of the North/South divide (Eckl and Weber 2007; Slater 1997; Therien 1999; Williams 1993, 2005). The gaps between the North and the South were understood to be narrowing, with countries graduating from the South to the North (Broad and Landi 1996), and the South was no longer seen to be poor or dependent on the North (Therien 1999). Ironically, these sorts of arguments were utilized by the US Bretton Woods institutions to reject the validity of the 1974 NIEO proposal, prescribing instead a more aggressive US-led neoliberal globalization project requiring structural adjustment and good governance (Slater 1997; Therien 1999). The need for the latter arose from the conviction that domestic factors such as corruption would negate efforts to alleviate poverty through structural transformation (Ould-Mey 2003). A circular argument appears to be at play here, with distributive justice claims thwarted on the basis of a rejection of the legitimacy of claims of North/South inequity, similar to the US negotiating strategy on climate change, as well as the color-blind approach in US racial politics (Omi and Winant 2015).

The Global North is also heterogeneous, and problems associated with the Global South are seen in the North, leading to the "Third Worldization of certain regions in the developed world" (McFarlane 2006; Toal 1994a, 231). The North/South configuration is therefore seen as a misleading interpretation of global inequality, with capitalism seen to be a more compelling framework (Wainwright 2010). Rather than use a binary metric of an over-consuming North and under-consuming South, Conca (2001, 68) suggested paying attention to an emergent "planetary middle class". Along similar lines, others problematize the state–centrism inherent in a North/South framing, questioning the validity of categorizing GHG emissions as well as impacts of climate change based on countries as opposed to corporations and communities (Barnett 2007, 1363). Thus the geopolitics of climate change is seen to be severely constrained by a 'territorial trap' that serves to privilege states as 'climate hegemons' – even as their regulatory power is declining (Agnew and Corbridge 1995; Conca 2001; O'Brien and Leichenko 2003). By privileging the inter-national scale, it is claimed, the disparities within both the Global North and the Global South are minimized, which is particularly damaging for vulnerable groups within developed countries (O'Brien and Leichenko 2003; Liverman 2009).

This is an important critique and reminder against the tendency to think of states as monoliths and to remain cognizant of disparities within states, but I believe that a both/an approach is preferable to an either/or approach in thinking about global inequality. In the context of climate change, there are relative winners and losers, and these exist "at all levels, from individuals to regions, nations, or groups of nations (i.e., advanced countries/developing countries)" (O'Brien and Leichenko 2003, 94). Since the determination of winners and losers from climate change is contingent upon scale of analysis, the choice of one scale over another becomes a political issue. The motivations for the self-identification of larger developing countries as victims have been questioned (DeSombre 2002; O'Brien and Leichenko 2003), with some cautioning against the prospect of an unholy alliance among the elites of the North and South (Barnett 2007; Norberg-Hodge 2008). There have been efforts to disaggregate emissions data for intra-state categories for more granularity beyond the national scale, and it does not seem necessary to do this at the expense of inter-state comparisons (Ananthapadmanabhan, Srinivas, and Gopal 2007; Baer et al. 2009). Responding to questions of legitimacy of equity arguments from "large polluters like India", Narain (2019, xxvii) points out that "the rich of India emitted less than the poorest American and there was no comparison between them and the rich of that world".

The legitimacy of sovereignty claims of nation-states has been questioned in light of economic globalization and the global nature of the climate crisis (Agnew and Corbridge 1995; DeSombre 2002; O'Brien and Leichenko 2003), but it may still be premature to dismiss the need to discuss climate mitigation responsibilities at the level of states. At least in theory, states have obligations to protect citizens' civil rights and the provision of social and environmental conditions that enable individual freedoms (Pellow 2007), and the imposition of additional pressures emanating from global environmental change affects their ability to carry out these obligations. While "the precise role of states in both mitigating and adapting to global environmental change is still in question", an outright dismissal of its relevance is premature at best (Biergmann and Dingwerth 2004, 12). There may be valid reasons to believe that political and other representatives of states could be corrupt and may manipulate the rhetoric of a

North/South divide, but to assume that to the effect of dismissing the relevance of the state in global environmental affairs is problematic, especially since negotiations under the UNFCCC occur between representatives of nation-states (DeSombre 2002).

Beyond issues of state-centrism, critical geopolitical scholars have problematized the North/South binary by drawing attention to identity politics: "a critical geopolitics is one that refuses the spatial topography of First World and Third World, North and South, state and state . . . we live in a condition of geopolitical vertigo" (Toal 1994b, 231). The task of critical geopolitics then is to highlight "the precariousness of these perspectival identities and the increasing rarefaction of geopolitical identities" (ibid). Yet the geopolitical vertigo Toal spoke of could itself be the product of a largely Euro-American centric worldview (Dodds 1998a; Joshi 2015). Others have argued that a North/South framing serves to reify the unequal power differentials between North and South through the implicit acceptance of the North's superiority, and consequently by condoning the North's paternalistic and interventionist role towards developing countries in international agreements and in advancing the project of modernity (Eckl and Weber 2007; Escobar 2004; Massey 2001; Slater 1997). Despite these critiques, the North/South geopolitical imagination continued to be regarded as a useful construct in environment-development scholarship, serving as a reminder of the linkages between geopolitics and development, of abiding core-periphery dynamics, and of imperialism and racism that remain as legacies of colonialism (Ould-Mey 2003; Potter 2001; Power 2006; Simon and Dodds 1998; Slater 2006). Contemporary scholars have continued to refer to inequality along a North/South divide. Many do so implicitly even as they critique these binaries (Williams 2005), a phenomenon Jones (2009, 176) termed the "paradox of categories".

In response to the issue of heterogeneity, some scholars have argued that the South or the Third World should not be seen as a monolith but rather as a diverse entity that fluctuates between acting in unity and maintaining plurality according to the geo-political context (Broad and Landi 1996; Williams 2005). In defending the term 'geo-politics' against charges of meaninglessness and polysemy, Toal (1994b, 260) had argued that implicit in such charges is a troubling assumption that "words and concepts have stable, assured identities which refer unproblematically and unambiguously to a fixed set of referents", urging for geo-politics to be seen not as "a singular, all-encompassing meaning or identity" (ibid, 269). Such courtesy should be offered to the idea of the Global South as well. While some countries of the South may have graduated to the North, such graduation is only partial – usually economic (Hansen 1980) – the self-identification of nations as members of the South owes to "a sense of shared vulnerability and a shared distrust of the prevailing world order rather than a common ordeal of poverty" (Najam 2004, 128). The South, for Anand (2004), represents the common experiences of people who have been victimized by a colonial and imperial past, and the similarities within the North and South outweigh their internal differences. This legacy has not only left countries of the South economically weaker and more vulnerable to the vagaries of a globalized capitalist economy (Williams 2005) but ensured its continued subjugation through an unequal international system where the South's voice wields less influence (O'Brien and Leichenko 2003; Therien 1999). The Southern bloc or the Third World coalition therefore makes sense when seen in the context of the dominance of industrialized states in global diplomacy and politics, as the coalition has offered developing countries with relatively marginal influence increased leverage in global negotiations (Hansen 1980; Williams 2005).

Proposals for structural reforms in the international economic system that emanated from the South have typically been met with tremendous resistance from Northern governments and institutions. Such proposals have either been rejected on the grounds that the South is not a valid category, or answered with palliative and status-quo-preserving measures such as the giving of aid for basic human needs for poverty alleviation (Guthman 1997), or countered with attempts at co-optation of the South's strongest states through graduation or integration. These states' resistance to such attempts implies a group identity within the South despite its heterogeneity (Therien 1999). Questioning the validity of the South as a meaningful category does little to change the status quo of the current asymmetric world order. An important question that arises then is to what extent doing so makes academics complicit in perpetuating the status quo. An ahistorical representation of territorial space and the invisibilization of ethical concerns in international relations seem to have favored the perpetuation of these unequal power relations into the present day (Agnew and Corbridge 1995; Therien 1999; Vogler 2016).

Given that the way we represent space is "deeply symbolic of how we define what is right and wrong and whom we identify with and against" (Agnew and Corbridge 1995, 79), the North/South geopolitical imagination appears to be a necessary intervention as it elicits a historical memory and an ethical dimension to international relations (Power 2006; Slater 1997). Toal (1994a, 230) urged geographers to read such a geopolitical imagination as a Lacanian imaginary, one that is "truly imaginary" with a significant divide between the real and the imagined. Toal's conviction that the Third World was devoid of any reality or material aspect admittedly drew upon a Western "imaginary ego-image" of the Third World. As such, it privileges a Western panoptic view of the Third World and effectively silences alternative understandings of the concept. How would discourses about the Third World or Global South look when the gaze considered is that of Third World subjects? And if these subjects constitute the so-called elites, would their perspectives be necessarily invalid?

An aspect of critiquing the North/South framework from a critical geopolitics approach has been to problematize the Third World elite, in particular by highlighting their manipulation of identities and political concepts. The term 'Third Worldism' has been used to signify such ideological posturing (Berger 2004). Here the implication is that identification with Third World solidarity is rhetorical, a charge others argue has yet to be verified (Roberts and Parks 2007; Williams 2005). Further, such a problematization has been typically based on a homogenizing and caricature of this 'elite' entity. Research on the agency of Third World or Southern subjects has been limited (Dodds 2001; Williams 2005). Analysis of the imaginary has instead been constrained by economic determinism, confined to structuralist accounts of the interests of these states. As such, changes in the global security and economic structures are seen to undermine the basis of Third World unity (Williams 2005). Williams (2005, 53) argued that the Third World or Global South is not a fixed category that represents some structurally determined political solidarity. Rather, it is an "imagined community of the powerless and vulnerable".

Critical geopolitics scholars have suggested the need to examine the reproduction of such imagined communities through research on "popular and elite forms of geopolitical reasoning" (Dodds 2001, 473). Thinking about categories not as "pre-given things-in-the-world but, rather, the result of [the] contingent and ongoing process of linking up locations of difference" helps us see geopolitical imaginaries such as the Global South or Third World as complex and dynamic rather than static and fixed, and that are continually renegotiated, rearticulated, and reproduced by individuals (Jones 2009, 180). It is individuals – activists, scholars and policy-makers – engaged in policy negotiations and discourses that substantiate a binary distinction among states as developed or developing, North or South, for a variety of reasons. Placing such agency on individuals helps us see these categories as subjectively produced rather than as objective fact. Given the power politics inherent in the process of categorization (Jones 2009), examining why these categories are contested by some and defended by others may be equally, if not more, important than analyzing their validity.

The spatial politics of climate mitigation

The politics of climate change are complex given the uneven distribution over space and time of the causes and consequences of climate change, as well as the capacity to deal with them. In the North/South framing of climate justice, the North is seen to have a historical responsibility towards the South for much of the cumulative greenhouse gases in the global atmospheric commons (Agarwal, Narain, and Sharma 1999; Desombre 2002). Yet, the politics of the contestation and reproduction of these categories suggest that these claims cannot be taken as a given. Nevertheless, neither can they be dismissed as illegitimate. What is certain in this analysis is that the spatial imaginary of the Global South has been enduring for its members – even with the erasure of the Annex I/non-Annex I binary – and as such has done the work of upholding the relevance of a concept of power, difference, and privilege at the scale of such a 'global regionalism' (Dodds 1998b). Fissures within G-77 and the fact that some of the G-77 countries have joined the G-7 to form the G-20 are offered as evidence to invalidate the viability of the negotiating bloc. Yet, an abiding sense of solidarity persists due to the inadequacy of Global North accountability for climate change, the prevailing sense of North/South inequities, as well as of mutual dependencies within the G-77.

The formation of the Third World Project and the Non-Aligned Movement, institutionalized in the G-77 from the start, sought to challenge the hegemony of colonial and imperial powers currently represented by the Global North. After an unsuccessful attempt in the 1970s, the politics of the NIEO seemed to be reinvigorated in the space of UNFCCC negotiations during the era of the Kyoto Protocol, 2005–20. In the movement from the Kyoto Protocol to the Paris Agreement, it appears that the United States has yet again succeeded in its ability to thwart Global South attempts for distributive and participatory justice and structural transformation towards more equitable global governance of the commons. Contestations of spatial categories were instrumental in these battles. The North/South distinction in the context of climate discourse serves as a spatial and scalar imaginary that represents a distinct historically informed claim to justice, institutionalized as the Annex I/non-Annex I dichotomy. Unsurprisingly, such claims have been met with counterarguments based on an invalidation of what this imaginary represents for countries disadvantaged in a globalized political economy. As I have indicated, there are interesting parallels between US responses to BIPOC communities and to the Third World – represented by rejection of the NIEO, the Kyoto Protocol, and minimization of racism in the United States – that would be a compelling area for further study. What would it mean for the core-periphery structure of the international system to be understood as a form of institutionalized or structural racism?

Even as the North/South frame for climate justice remains vital, questions raised to challenge its validity should not be dismissed. While championing North/South equity, its advocates should be answerable to charges of state-centrism, and elite manipulation of the Global South imaginary for geopolitical power. The challenge of attending to questions of the Global South within the Global North can also not be ignored, as also those seeking racial and economic justice within the United States cannot in good conscience remain oblivious to the depredations of the United States in global context. There are competing claims of reparations from the Global South within and without the Global North. How should the relative urgency of either set of claims be weighed? It is also worth asking how relationships of transnational solidarity among the marginalized may be built across the North-South divide; and relatedly, if those marginalized in the Global North and South are to be deemed to be part of the spatial imaginary of the Global South, how will their relationships with their respective nation-state configurations – some hegemonic and others marginalized in the global system – affect such transnational solidarity? I believe these questions are important for a pursuit of climate justice that is not limited to US-centrism or the centering of the European experience, which are implicit ways of exercising state-centrism.

While North/South climate politics have centered on states as legitimate arbiters of responsibility and vulnerability, such state-centrism has been challenged both on the basis of the inaccuracy of aggregating emissions by state and on the basis of the state's inability to represent the interests of the most climate-vulnerable populations. Calls have therefore been made to eschew such a territorial trap for a "more empowering and critical geopolitics" that takes a "more subaltern and class-based view" that places the burden of action on the political and economic choices of people in developed countries (Barnett 2007, 1372) or places hope in transnational environmental and social movements (Escobar 2004). Rather than calling for a disavowal of the international system, my sense is that a climate governance architecture that is polycentric in ways that maintain the viability of states while creating spaces and institutions for meaningful representations of those who are marginalized within the state, including women, youth, Indigenous groups, racialized or ethnic groups, the unhoused, differently abled, and so on. Individuals are racialized, gendered, classed in particular ways that shape their vulnerability to disasters, but they are also in the contemporary context always tethered to a nation-state – whether as a citizen, an immigrant, or refugee – with whom relationships of entitlement and obligation are established.

Notwithstanding claims that globalization has weakened the role of the state, these relationships remain and merit consideration in the context of climate justice claims. COVID-19 has showed us how response to a worldwide disaster can depend on the character of state intervention. I argue that it is too soon to eschew state-centric examinations of climate inequities and injustice because pervasive difference and inequities exist at this scale and because arguments about Global South heterogeneity have much too often been used to invalidate claims for justice. I question the assumption that a critical geopolitics of climate change must be defined by a critique of state-centrism or geopolitical imaginaries of Global South and North in order to be 'critical' and suggest that states can and should be expected to play a more enabling role in meeting the needs of their most vulnerable citizens. In a climate of neoliberalism, arguments invalidating the authority of the state turn easily into circular arguments where delegitimization of justice claims serves as both rationale and outcome.

References

Agarwal, Anil, Sunita Narain, and Anju Sharma, eds. 1999. *Green Politics*. New Delhi: Center for Science and Environment.

Agnew, John, and Stuart Corbridge. 1995. "The Territorial Trap." In *Mastering Space: Hegemony, Territory and International Political Economy*, edited by John Agnew, 78–100. London: Routledge.

Anand, Ruchi. 2004. *International Environmental Justice: A North/South Dimension*. Burlington: Ashgate.

Ananthapadmanabhan, G., K. Srinivas, and Vinuta Gopal. 2007. *Hiding Behind the Poor: A Report by Greenpeace on Climate Injustice – An Indian Perspective*. Bangalore: Greenpeace India Society.

Baer, Paul, Tom Athanasiou, Sivan Kartha, and Eric Kemp-Benedict. 2009. "Greenhouse Development Rights: A Proposal for a Fair Global Climate Treaty." *Ethics, Policy & Environment* 12, no. 3: 267–81.

Barnett, Jon. 2007. "The Geopolitics of Climate Change." *Geography Compass* 1, no. 6: 1361–75.

Berger, Mark T. 2004. "After the Third World? History, Destiny and the Fate of Third Worldism." *Third World Quarterly* 25, no. 1: 9–39.

Bhagwati, Jagdish N., ed. 1977. *The New International Economic Order.* Cambridge, MA: The MIT Press.

Biergmann, Frank, and Klaus Dingwerth. 2004. "Global Environmental Change and the Nation-State." *Global Environmental Politics* 4, no. 1: 1–22.

Bose, Sugata, and Ayesha Jalal. 2011. *Modern South Asia*, 3rd ed. New York: Routledge.

Broad, Robin, and Christina M. Landi. 1996. "Whither the North/South Gap?" *Third World Quarterly* 17, no. 1: 7–17.

Castro, João A. D. A. 1972. "Environment and Development: The Case of Developing Countries." In *Green Planet Blues, 2010*, edited by Ken Conca and Geoffrey D. Dabelko. Boulder: Westview Press.

Conca, Ken. 2001. "Consumption and Environment in a Global Economy." *Global Environmental Politics* 1, no. 3: 53–71.

DeSombre, Elizabeth R. 2002. *The Global Environment and World Politics.* New York: Continuum.

Dodds, Klaus. 1998a. "Book Review: Critical Geopolitics." *Economic Geography* 74, no. 1: 77–79.

———. 1998b. "The Geopolitics of Regionalism." *Third World Quarterly* 19, no. 4: 725–43.

———. 2001. "Political Geography III: Critical Geopolitics After Ten Years." *Progress in Human Geography* 25, no. 3: 469–84.

Dubash, Navroz K. 2009. *Toward a Progressive Indian and Global Climate Politics.* New Delhi: Centre for Policy Research.

Eckl, Julian, and Ralph Weber. 2007. "North/South? Pitfalls of Dividing the World by Words." *Third World Quarterly* 28, no. 1: 3–23.

Escobar, Arturo. 2004. "Beyond the Third World: Imperial Globality, Global Coloniality and Anti-Globalisation Social Movements." *Third World Quarterly* 25, no. 1: 207–30.

Gilman, Nils. 2015. "The New International Economic Order: A Reintroduction." *Humanity* 6, no. 1: 1–16.

Government of India. 2008. "National Action Plan on Climate Change: Prime Minister's Council on Climate Change." https://www.ncbi.nlm.nih.gov/pmc/articles/PMC2822162/.

Gupta, Himangana, Ravinder K. Kohli, and Amrik S. Ahluwalia. 2015. "Mapping 'Consistency' in India's Climate Change Position: Dynamics and Dilemmas of Science Diplomacy." *Ambio* 44, no. 6: 592–99.

Guthman, Julie. 1997. "Representing Crisis: The Theory of Himalayan Environmental Degradation and the Project of Development in Post-Rana Nepal." *Development and Change* 28: 45–69.

Hansen, Roger D. 1980. "North/South Policy – What's the Problem?" *Foreign Affairs* 58, no. 5: 1104–28.

India. n.d. "INDC: India's Intended Nationally Determined Contribution: Working Towards Climate Justice." www.indiaenvironmentportal.org.in/content/419700/indias-intended-nationally-determined-contribution-working-towards-climate-justice/.

Isbister, John. 2006. *Promises Not Kept*, 7th ed. Bloomfield: Kumarian Press.

Jakobsen, Susanne F. 1998. *India's Position on Climate Change from Rio to Kyoto.* Working Paper 98.11. Copenhagen: Centre for Development Research.

Jha, Laxmi K. 1982. *North South Debate.* New Delhi: Chanakya Publications.

Jones, Reece. 2009. "Categories, Borders and Boundaries." *Progress in Human Geography* 33, no. 2: 174–89.

Joshi, Shangrila. 2013. "Understanding India's Representation of North – South Climate Politics." *Global Environmental Politics* 13, no. 2: 128–47.

———. 2014. "North/South Relations: Colonialism, Empire and International Order." In *Routledge Handbook of Global Environmental Politics*, edited by Paul G. Harris. New York: Routledge.

———. 2015. "Postcoloniality and the North/South Binary Revisited: The Case of India's Climate Politics." In *The International Handbook of Political Ecology*, edited by Raymond L. Bryant. Cheltenham: Edward Elgar.

Liverman, Diana. 2009. "The Geopolitics of Climate Change: Avoiding Determinism, Fostering Sustainable Development. An Editorial Comment." *Climatic Change* 96: 7–11.

Martinez-Alier, Joan. 2002. "Green Justice – Ecological Debt and Property Rights on Carbon Sinks and Reservoirs." *Capitalism Nature Socialism* 13, no. 1: 115–19.

Massey, Doreen. 2001. "Geography on the Agenda." *Progress in Human Geography* 25, no. 1: 5–17.

McFarlane, Colin. 2006. "Crossing Borders: Development, Learning and the North/South Divide." *Third World Quarterly* 27, no. 8: 1413–37.

Meadows, Donella H., Dennis L. Meadows, Jorden Randers, and William W. Behrens. 1972. *The Limits to Growth*. New York: Universe Books.

Miller, Marian. 1995. *The Third World in Global Environmental Politics*. Buckingham: Open University Press.

Nabhan, Gary P. 1998. "Cultural Parallax in Viewing North American Habitats." In *Environmental Ethics*, edited by Richard Botzler and Susan Armstrong, 2nd ed. New York: McGraw Hill.

Najam, Adil. 2004. "Dynamics of the Southern Collective: Developing Countries in Desertification Negotiations." *Global Environmental Politics* 4, no. 3: 128–54.

Najam, Adil, Saleemul Huq, and Youba Sokona. 2003. "Climate Negotiations Beyond Kyoto: Developing Countries Concerns and Interests." *Climate Policy* 3: 221–31.

Narain, Sunita. 2019. "Equity: The Final Frontier for an Effective Climate Change Agreement." In *Climate Futures*, edited by Kum-Kum Bhavnani, John Foran, Priya A. Kurian, and Debashish Munshi. London: Zed Books.

Norberg-Hodge, Helena. 2008. "The North/South Divide." *Ecologist* 38, no. 2: 14–15.

O'Brien, Karen L., and Robin M. Leichenko. 2003. "Winners and Losers in the Context of Global Change." *Annals of the Association of American Geographers* 93, no. 1: 89–103.

Omi, Michael, and Howard Winant. 2015. *Racial Formation in the United States*, 3rd ed. New York: Routledge.

Ould-Mey, Mohameden 2003. "Currency Devaluation and Resource Transfer from the South to the North." *Annals of the Association of American Geographers* 93, no. 2: 463–84.

Paterson, Matthew. 1996. *Global Warming and Global Politics*. New York: Routledge.

Pellow, David N. 2007. *Resisting Global Toxics*. Cambridge, MA: The MIT Press.

Potter, Rob. 2001. "What Ever Happened to Development Geography?" *The Geographical Journal* 167, no. 2: 188–89.

Power, Marcus. 2006. "Anti-Racism, Deconstruction and 'Overdevelopment'." *Progress in Development Studies* 6, no. 1: 24–39.

Prashad, Vijay. 2014. *The Poorer Nations*. New York: Verso.

Rajan, Mukund G. 1997. *Global Environmental Politics*. New Delhi: Oxford University Press.

Roberts, J. Timmons, and B. C. Parks. 2007. *A Climate of Injustice*. Cambridge, MA: The MIT Press.

Simon, David, and Klaus Dodds. 1998. "Introduction: Rethinking Geographies of North/South Development." *Third World Quarterly* 19, no. 4: 595–606.

Slater, David. 1997. "Geopolitical Imaginations Across the North/South Divide: Issues of Difference, Development and Power." *Political Geography* 16, no. 8: 631–53.

———. 2006. "Imperial Powers and Democratic Imaginations." *Third World Quarterly* 27 no. 8: 1369–86.

Sowers, Jeannie. 2007. "The Many Injustices of Climate Change." *Global Environmental Politics* 7, no. 4: 140–46.

Srinivasan, U. Thara et al. 2008. "The Debt of Nations and the Distribution of Ecological Impacts from Human Activities." *Proceedings of the National Academy of Sciences* 105 no. 5: 1768–73.

Taylor, Marcus. 2014. *The Political Ecology of Climate Change Adaptation*. New York: Routledge.

Therien, Jean-Philippe. 1999. "Beyond the North/South Divide: The Two Tales of World Poverty." *Third World Quarterly* 20, no. 4: 723–42.

Toal, Gerard. 1994a. "Critical Geopolitics and Development Theory: Intensifying the Dialogue." *Transactions of the Institute of British Geographers* 19, no. 2: 228–33.

———. 1994b. "Problematizing Geopolitics: Survey, Statesmanship and Strategy." *Transactions of the Institute of British Geographers* 19, no. 3: 259–72.

United Nations. 1992. *United Nations Framework Convention on Climate Change*, chapter XXVII, Environment, title 7. New York: United Nations, May 9. https://treaties.un.org/pages/ViewDetailsIII.aspx?src=TREATY&mtdsg_no=XXVII-7&chapter=27&Temp=mtdsg3&clang=_en.

———. 1998. "Kyoto Protocol to the United Nations Framework Convention on Climate Change." https://enb.iisd.org/process/climate_atm-fcccintro.html.

———. 2015. *Paris Agreement*, chapter XXVII, Environment, title 7d. Paris: United Nations, December 12.

Vidal, John. 2010. "Confidential Document Reveals Obama's Hardline US Climate Talk Strategy." *The Guardian*, April 12.

Vihma, Antto. 2010. *Elephant in the Room: The New G77 and China Dynamics in Climate Talks*. FIIA Briefing Paper 62. Helsinki: Finnish Institute of International Affairs, May 26. https://www.fiia.fi/en/publication/elephant-in-the-room.

———. 2011. "India and the Global Climate Governance: Between Principles and Pragmatism." *Journal of Environment & Development* 20, no. 1: 69–94.

Vogler, John. 2016. *Climate Change in World Politics*. New York: Palgrave Macmillan.

Wainwright, Joel. 2010. "Climate Change, Capitalism, and the Challenge of Transdisciplinarity." *Annals of the Association of American Geographers* 100, no. 4: 983–91.

Williams, Marc. 1993. "Re-Articulating the Third World Coalition: The Role of the Environmental Agenda." *Third World Quarterly* 14, no. 1: 7–29.

———. 2005. "The Third World and Global Environmental Negotiations: Interests, Institutions and Ideas." *Global Environmental Politics* 5, no. 3: 48–69.

3 Postcolonialism and the struggle over the atmospheric commons

Discourse about colonialism and decolonization is often centered on the European variants for good reason – the scope of the European colonial project was global. The abiding legacies on language hegemony and epistemology are unmistakable. Yet variations of the colonial project at more intimate sites and scales are no less real to those affected. For a Newar from the Kathmandu Valley – known as Nepa by the Newars – self-identification as a Nepali citizen is itself a marker of colonization. While on the one hand, Prithvi Narayan Shah, the ruler of Gorkha who created the nation-state of Nepal in 1768–69, did so by 'unifying' the many erstwhile Himalayan kingdoms in order to successfully withstand encroachment by the British East India Company, it did so by defeating the Newar kingdoms of the Kathmandu Valley in a brutal war that would be followed by the colonial subjugation of the Newar people, including the suppression of their language, culture, and their resource commons. Two and a half centuries later Nepal is a Federal Republic with multiculturalism and equality codified in the 2015 Constitution, but the hegemony of the Khas/Arya ethnicity and Nepali language persists in institutionalized forms in bureaucratic structures of governance. Assaults on resource governance structures such as the *Guthi* – deemed central to Newar identity and a civilization that is over 2000 years old – continue in the present day, with ongoing struggles in 2020. For an Indigenous person in South Asia, struggles with the modern nation-state are not uncommon (Roy 2004; Sivaramakrishnan and Cederlof 2006). *Jal, Jangal, Zameen* is the rallying cry of the Gonds of Telangana and Jharkhandi *Adivasis*, among many others, against the Indian nation-state in response to its decades-long encroachment on *Adivasi* sovereignty and self-determination (Damodaran 2006; Parenti 2011). Both the Indian and the Nepalese nation-states exercise forms of colonialism over their own citizens in these ways. But there is also the bilateral relationship between the two countries that is noteworthy.

Although Nepal is an older nation-state than India and they have much shared cultural, ethnic, and linguistic heritage between them, the power dynamic between the two countries has always been uneven – in no small part due to the history of British colonial rule – India has often assumed a paternalistic and patronizing position towards Nepal. This was reflected most recently following India's revoking of Jammu and Kashmir's constitutional autonomy in late 2019 when a new map publicized by India renewed territorial disputes with Nepal that have been attributed to "cartographic manipulation with a sinister motive" during the British colonial era, and perceived as a "blatant violation of Nepali sovereignty in total contravention of international norms" (Cowan 2015; Manandhar and Koirala 2001, 3–4; Zehra 2019). Gorkha-ruled Nepal in the 18th century sought to extend its dominion and territory beyond what its present-day borders are, but receded much of it in a 1915–16 treaty called the *Sugauli Sandhi* to stop war with the British East India Company. In the treaty, Nepal and British India agreed that the Kali river would form the border between the two sovereign powers. In maps produced between 1816 and 1856 the headwaters of the Kali river were recognized to be in Limpiyadhura, but cartographic manipulations made during the years 1857–75 would claim the headwaters of the Kali River to be southeast of the original headwaters, in Lipulekh. The updated map published by India in 2019 exercised further territorial transgression in disputed territory to legitimize Indian jurisdiction over a strategic trade and pilgrimage route between India and Tibet called Lipulekh Pass, prompting Nepal to assert sovereignty with its own updated map (Bose 2020; Manandhar and Koirala 2001).

While a full analysis of this recent bilateral dispute is beyond the scope of this text, I bring it up here, firstly, to highlight that the Global South is far from a monolith, and, secondly, to indicate that North/South struggles over the atmospheric commons cannot be divorced from struggles over land within the Global South. There exist sharp disparities within India itself and within the different nation-states that self-identify as nations of the Global South. India is clearly one of the more economically and politically powerful countries within the Global South (Rothermund 2008). It is fair to say that India is a regional hegemon in South Asia. But I bring up the textured depiction of (my own) identity in relation to different forms of colonialism earlier to illustrate the centrality of the nation-state scale of reference in contemporary politics – its reification despite its limitations – and to emphasize the importance of a two-way politics of scale. The fluidity and multi-faceted nature of identity enables a politics of scale spanning the individual to national to global and others in between. These claims need not be mutually exclusive. Despite the heterogeneity of the Global South and what is contained therein, this spatial imaginary resonates as a powerful counterhegemonic notion against Euro-centric colonial power dynamics. The Newar struggle against Nepali dominance and the Nepali struggle against Indian or Chinese dominance can simultaneously coexist with the joint struggle of various groups in Nepal, India, China, and many other countries against the hegemony of the United States and other countries in the Global North. Although the Global North and Global South are not inherently natural formations, they have come to be reified in the context of international relations and exchange. Debates over these identity markers are to be expected, and seeking to understand why these debates are occurring and what their implications are is necessary.

Postcolonial theory and political ecology

A widely cited article titled 'Global Warming in an Unequal World' published by the CSE in Delhi pre-figured the North/South climate negotiations of the following decades and unequivocally placed it in the context of colonialism:

> the idea that developing countries like India and China must share the blame for heating up the earth and destabilizing its climate, as espoused in a recent study published in the United States by the World Resources Institute in collaboration with the United Nations, is an excellent example of environmental colonialism.
>
> (Agarwal and Narain 1990, 1)

The authors pointed out that "the gargantuan consumption of the developed countries, particularly the United States" was mostly responsible for climate change. Yet, instead of seeking to address this hyperconsumption, they lamented, Western environmentalists were focusing on the potential increase in consumption of the average Indian and Chinese citizen. Denouncing the "highly partisan 'one worldism'" inherent in Western prescriptions inspired by discourses of "Our Common Future" and intergenerational justice, they urged environmentalists in the Third World to ask those in the West, "whose

Figure 3.1 Mural at the JNU campus, Delhi, December 2008

future generations are we seeking to protect, the Western World's or the Third World's?" (ibid, 18). These questions remain relevant in the context of seeking to understand the global dimensions of climate justice, despite and precisely because of the movement from the Kyoto Protocol to the Paris Agreement and the consequent erasure therein of North/South difference that were codified as Annex I and non-Annex I parties in the Kyoto Protocol.

The Global South in postcolonial theory

Words have particular meanings within particular communities of practice and in particular spatial and temporal contexts. Meanings are never fixed and are rather continually interrogated and sometimes reclaimed. Such is the case with the term 'Global South' as I have argued and review here (Joshi 2015). In the body of work recognized as postcolonial theory, terms such as 'Third World' and 'Global South' have sometimes been associated with a pejorative connotation, which I will unpack here. But first, a note about the term 'postcolonial', which itself has been deemed problematic in its presumption of a world that has already undergone successful decolonization, and in this conceptualization either negates the experience of settler colonialism in places such as the United States or perpetuates the "myth of a 'postcolonial' world" based merely on 'juridical-political decolonization' when a condition of coloniality persists (Grosfoguel 2011, 14). Indeed, the term 'postcolonial' appears to be often interpreted as a period after colonization, marked by political independence from colonial rule. But postcolonial scholars such as Gandhi (1998) and Said (1994) have used a hyphenated 'post-colonial' to represent that particular condition while using the term 'postcolonial' to refer to a condition that – not unlike coloniality – continues to be marked by "the colonial aftermath" (Gandhi 1998, 4). As such the understanding is that colonialism has fundamentally altered the world; the postcolonial project seeks to grapple with that aftermath. Although postcolonial theory is a body of work originating from various post-colonial locations, its insights are relevant in settler colonial contexts as well. In the post-colonial context, the legacies of colonialism are understood to remain, "in new forms of domination that follow and extend old imperial lines of unequal interconnection" (Nash 2004, 105). Neocolonialism thus refers to "forms of political and economic domination through which the West continues to exploit much of the world" (ibid, 113) through a set of "political, ideological, economic, and social practices" (Said 1994, 9). It follows that central to postcolonial theorizing is "a critical engagement with colonialism and its continued legacies" (Nash 2004, 105).

In postcolonial theory, oppositional representations of the world along lines of First World/Third World, East/West, North/South, developed/underdeveloped, core/periphery have been critiqued as colonial binaries that assume the inferiority of the Third World and the Global South (Nash 2004). This line of critique draws on Said's (1978) treatise on Orientalism, although Said (1994, 52) had himself gone on to remark that: "no identity can ever exist by itself and without an array of opposites, negatives, oppositions". He had also argued that the colonizer-colonized relationship had reemerged in the similarly 'compartmentalized' reincarnation of "what is often referred to as the North/South relationship" Said (1994, 17). While this statement is open to interpretation, a postcolonial approach to geography went on to be synonymous with disavowing the North/South framing in geopolitics and international relations. Nash (2004, 110) argued that these constructions "draw on colonial traditions of representation" of 'us' and 'them'. Such othering practices were deemed to be responsible for obscuring the role of Western imperialism in subjugating the Third World by naturalizing the superiority and the success of the West over the rest of the world (Sidaway 2002). Further, a state-centric North/South divide was deemed problematic for concealing dynamics of 'internal colonialism' based on hierarchies of power in the postcolonial state and that of 'ultraimperialism' that manifests beyond nationalist tropes and through multinational capital as represented by the International Monetary Fund and the World Bank (WB) among others (Blunt and McEwan 2002; Sidaway 2000, 2002). Eventually, postcolonialism in geography would challenge "the binary categories of homogenous colonizing and colonized groups" (Nash 2004, 124). It should not be too surprising, then, that critical geography would be largely silent on North/South politics, save the exceptions of Simon and Dodds (1998), Slater (2004), and Power (2006), even as research *in* the Global South by scholars from the North has flourished.

In much of the geographical research in the Global South using postcolonial approaches, a common practice has been to critique Western theories of modernist development harkening back to the 1950s and 1960s (McEwan 2003; Nash 2004). The foundational basis for such a development theory is understood to be "Enlightenment ideals of modernity and progress, and . . . colonial discourses of the 'civilising mission'" that accompanied paternalistic attitudes and interventions. These discourses are often seen as nothing more than "discursive tools that justify the neoliberal march of free market capitalism" (Nash 2004, 110). Extending this logic, McEwan (2003, 343) claimed that postcolonial theory had not paid enough attention to global capitalism and class analysis such that its "politics of recognition" did not sufficiently extend to a "politics of distribution" needed to address influence global power disparities. A related assertion was that postcolonial theorizing tended to dwell on the discursive and not enough on the materialities of postcolonialism. A critical geographical approach to postcolonial theorizing, it was argued, could fill these deficits in a predominantly textual and culturally oriented field of study. In this vein, the framing 'postcolonial geographies' was meant to signify the myriad ways in which the legacies of colonialism unfold within the Global South, with postcolonialism understood to be a "geographically dispersed contestation of colonial power and knowledge", while the issue of North/South power symmetries remained sidelined (Blunt and McEwan 2002, 4). Implicated in both sets of arguments invalidating or invisibilizing the North/South axis of difference is the centrality of a class-based analysis of capitalism and the consequent problematization of elites in postcolonial locations as a key obstacle to eliminating global inequalities. If postcolonial geography is to be decolonized (Grosfoguel 2011), the North/South dimensions of coloniality or postcolonialism cannot continue to be invisibilized, while the Global South continues to serve merely as an intellectual playground for First World academics to conduct marginality studies (Gandhi 1998).

A North/South postcolonial geography?

I argue that postcolonial theory offers us ways in which to read the North/South relationship in a different manner than what the preceding discussion reveals. I draw on insights from Bhabha, Grosfoguel, Said, Spivak, and others who have used their thinking in geography and cognate disciplines to make my argument about the abiding relevance of a North/South framing for postcolonial climate politics (Joshi 2015). Notions of strategic essentialism, hybridity, and reflexivity are enabling; and tendencies towards Orientalism, Eurocentrism, and complicity ought to give us pause.

Critique and complicity

Almost three decades ago, Jane Jacobs posed what is considered "one of the most significant questions for contemporary human geography" (Nash 2004, 105): "Can the spatial discipline of geography move from its positioning of colonial complicity towards producing postcolonial spatial narratives?" (Jacobs 1996, 163). The risk of academic complicity with reinforcing global inequalities is a serious one, and anyone committed to using their scholarship to enhance social justice rather than maintaining the status quo would be wary of it. Postcolonial scholars (Kapoor 2008; Spivak 2010a) have warned of such complicity and suggested that attention to reflexivity is crucial in countering such complicity. I take this to mean that a scholarly examination using a postcolonial perspective takes seriously into consideration the author's own positionality, based on both social and epistemic location (Grosfoguel 2011). This must happen if decolonial work on postcolonialism is to be more than an intellectual exercise.

The emphasis on postcolonial critique in the form of the contestation of 'colonial' binary identities is understandable given the role they have played in accentuating colonial power (Jacobs 1996). However, there is an aspect to this power and identity that often goes underexamined – an identity politics based on the appropriation of binary construction as anti-colonial strategy. Postcolonial scholars have drawn attention to decolonial strategies that utilize the "disruptive power of hybridity" employing the colonizer's tools of oppression (Jacobs 1996, 14; Kapoor 2008) in a way that achieves 'subversive complicity' (Grosfoguel 2011, 24). Jacobs (1996, 14) thus spoke of the ability of colonized groups to subvert colonial power "through disruptive inhabitations of colonialist constructs . . . [enabled by the] . . . vulnerability of imperialist and colonialist power . . . against anticolonial formations". The colonizer's negative constructions of the colonized other are often appropriated in counter-colonial efforts:

> The processes by which notions of the Self and Other are defined, articulated and negotiated are a crucial part of what might be thought of as the cultural dimension of [both] colonialism and postcolonialisms. [This] making and remaking of identity occurs through representational and discursive spheres.
>
> (Jacobs 1996, 2)

Hybridity refers to the embodiment of a bicameral identity that serves to destabilize hegemonic authority (Kapoor 2008). "A contingent, border-line experience opens up *in-between* colonizer and colonized. This is a space of cultural and interpretive undecidability produced in the 'present' of the colonial moment" (Bhabha 1994, 295–96). Drawing on Bhabha's (1994, 296) "margin of hybridity" where cultural oppositions are momentarily suspended, Kapoor (2008, 139) deduces that a 'hybridizing strategy' can exploit the indeterminacies of power to co-opt the dominant modality to beat the colonizers at their own game. Such attempts are not necessarily guaranteed success because of the "ever-changing forms of neocolonial hegemony" (Kapoor 2008, 147) – the game keeps evolving. These resistance strategies are adopted in dialectical relation to the "tenacious and adaptive power" of colonial discourses that seek to continually reinvent and reinscribe the status quo (Jacobs 1996, 14). Awareness of these two-way power dynamics obligates postcolonial scholars "not to accept the politics of identity as given, but to show how all representations are constructed, for what purpose, by whom, and with what components" (Said 1994, 314). As Laclau (1990, 31) has said, "the constitution of a social identity is an act of power and that identity as such is power". Such an understanding of North/South identities as "social constructs, and strategic ones at that, destabilizes a whole range of claims" (Jacobs 1996, 162).

Postcolonialism's critique of North/South binaries has been predicated mostly on discourses from the colonizers' vantage point and as such have seemingly focused solely on the "continued legacies of colonialism [rather] than challenges to them" (Nash 2002, 221). But binary 'othering' is not an act that can be performed only by dominant groups. It is also a strategy of resistance to dominance/hegemony by the marginalized. Binaries help us make sense of differences in the world. To understand who is behind particular representational practices and for what purpose is therefore important. A neglected but important task is therefore to engage with the multiple forms of agency of formerly colonized people without dismissing them a priori as 'elitist' due to perceived closeness with the colonizer's image. The elite-poor binary often used to characterize politics within India as well as in its portrayal without has been found misleading:

> the dominant imagery of India's new middle class, exemplified by glitzy shopping malls and international travel, in fact conflates India's new (and tiny) globalized elite with the large and heterogeneous middle class, comprising a wide range of incomes and practices, within which elites are a small component.
>
> (Lemanski and Lama-Rewal 2013, 91)

There is a disturbing parallel in the way that Third World 'elites' and 'emerging economies' in the Global South are problematized by scholars – both are conjured up as the 'straw man' whose very existence invalidates their claims on behalf of their broader constituency (Joshi 2015).

The question of representation – can the 'elite' of a nation truly represent the interests of ordinary citizens in international relations; and can emerging economies truly represent the interests of the Global South – is reminiscent of Spivak's (2010a) charge of White saviorism of academics, because if the Third World 'elite' cannot be trusted to represent the Third World, by implication, then the one making the charge (presumably from a First World location or place of privilege) is presenting themselves as more trustworthy in representing the voice or concerns of the Third World. Such critiques, when performed unreflexively by Northern academics, render them complicit – coincidentally alongside imperialist strategists in hegemonic power centers – in the maintenance of the unequal status quo between the North and South and hence of coloniality (Grosfoguel 2011). While the contestations may occur in the discursive realm, the implications are for material resources through economic relations and environmental responsibilities. Anand's (2004) portrayal of the Global South as united by a shared experience of vulnerability in the world political economic system due to their victimization by a colonial and imperial past can be disavowed as a Third Worldist identity politics or as a legitimate appeal to a strategic essentialism (Spivak 1993) where claiming victimhood increases bargaining power.

The NIEO proposal was made on similar grounds (Bhagwati 1977), where a coalition of formerly colonized countries sought to challenge the unfair terms of trade between themselves and industrialized countries, by demanding changes that would enable the South to achieve self-sustaining economic growth and industrialization, although the Bretton Woods institutions rejected these calls by pointing to the fallacy of the North/South divide. The increasing salience of global environmental concerns opened up new possibilities for North/South diplomacy by enhancing the bargaining power of the South in global environmental politics (Anand 2004). But when Western scholars negate these claims to justice as 'ideological posturing by elites' (Berger 2004) or an 'uncritical' geopolitics (Barnett 2007; Toal 1994), they become unwittingly complicit in the preservation of the unequal North/South status quo, demonstrating "how the production of western knowledge is inseparable from the exercise of Western power" (Blunt and McEwan 2002, 6). For in denying the North/South dichotomy, these scholars (of the Global North) lend strength to the assertion – akin to that of the Bretton Woods institutions in the context of NIEO proposals – that more and less powerful countries do not exist, and that the material well-being of the poor in the Global South can be addressed without challenging the fundamental structures of the global political economy that are skewed towards the Global North.

Towards a postcolonial political ecology of the atmospheric commons

Political ecology has a long tradition of addressing the structural drivers of ecological crises and of doing so by situating specific issues within a historical context shaped by colonialism and capitalism (Blaikie and Brookfield 1987). Although the field incorporated postcolonial perspectives that sought to create space for "thinking about the complicity of academic narratives in the extension of colonial power and repression, even narratives that ostensibly represent emancipatory ideologies" (Robbins 2011, 69), with few exceptions, the explanatory powers of a postcolonial political ecology have not been utilized to seek to explore the 'inchoate possibilities' of the 'unfailure' that is NIEO. Laclau (1990, 31) describes 'inchoate possibilities' as "those whose actualization was once attempted but were cancelled out of existence". Gilman (2015, 10) describes an 'unfailure' as:

> the paradox that many seemingly failed political and social movements, even though they did not realize their ambitions in their own moment, often live on as prophetic visions, available as an idiom for future generations to articulate their own hopes and dreams. . . . The unfailed afterlife of the NIEO is perhaps most evident today in global climate change negotiations.

These unchoate possibilities of the NIEO present a salient case for a scaled-up postcolonial political ecology analysis of climate negotiations as the specifics of the Paris Agreement are deliberated on in the coming years. Two major concerns inherent in political ecology scholarship are central to this examination – a focus on the politics of access and control over resources that is cognizant of "the role of unequal power relations in constituting a politicized environment" (Bryant 1998, 79; Watts and Peet 2004), and a normative orientation towards redistributive justice and ecological sustainability that are derived from a "basic radical ethical position" (Bryant and Jarosz 2004, 808). A Gramscian approach to political ecology has been identified as particularly relevant because of its emphasis on the 'ethico-political' dimensions of scholarship (Ekers, Loftus, and Mann 2009; Mann 2009). Feminist geographers (Jarosz 2004), decolonial thinkers (Grosfoguel 2011), and postcolonial thinkers (Kapoor 2008; Spivak 2010a) have emphasized the importance of critical reflexivity of a scholar's positionality and epistemic location, and it is also present in Gramsci's thinking on Marxism (Gramsci 1957).

Early political ecology focused on the dynamics of environmental change in the Third World with an emphasis on problematizing neo-Malthusian explanations of environmental degradation (Blaikie and Brookfield 1987). Uneven access to, and power over, environmental resources and "the ways in which conflict over access to environmental resources is linked to systems of political and economic control first elaborated during the colonial era" (Bryant 1998, 79) have been key themes in the literature. There have been efforts to bring political ecology to the First World (Robbins and Sharp 2003) as well as to formulate regional political ecologies in both South (Blaikie and Brookfield 1987) and North (Walker 2003). While political ecology has judiciously sought to situate explanations of environmental change within the context of a globalized political economy, few researchers have sought to address North/South power dynamics in struggles over the atmospheric commons (Joshi 2015; Kim 2009). I have found the Gramscian notions of hegemony and counterhegemony to be helpful in scaling up political ecology to a North/South context (Cox 1993; Doty 1996). The global atmosphere serves as the environment over which hegemony is reinscribed or challenged (Ekers, Loftus, and Mann 2009). This can happen discursively: "the hegemonic dimension of global politics is inextricably linked to representational practices . . . how certain representations underlie the production of knowledge and identities and how these representations make various courses of action possible" (Doty 1996, 8); but of course the discursive is inherently connected to the material (Neumann 2009). Identity categories such as class, state elites, North/South are discursively constructed and contested in efforts to maintain or oppose hegemony and therefore should not be assumed a priori (Doty 1996). As Watts and Peet (2004, 25) also argued, power struggles over access to resources "are invariably wrapped up with questions of identity. . . . [T]hese forms of identity are not stable (their histories are often shallow), and may be put to use (they are interpreted and contested) by particular constituencies with particular interests."

Postcolonialism and development

A key aspect of postcolonial political ecology has been to problematize modernization and development as Enlightenment-based Western practices and systems of knowledge. Significantly, Bebbington (2004) and Rangan (2004) have shown, on the basis of their work in the Ecuadorian Andes and the Indian Himalaya respectively, that the binary distinction between local/Indigenous sovereignty and modernization/development is unhelpful. Rather than eschewing modern technologies, Indigenous farmers were more interested in increasing their control over their ability to bring about social change, prompting Bebbington (2004) to suggest that rather than leading to cultural erosion, modernization can sometimes become a means of cultural survival. Rather than being a "hegemonic discourse of the West", Rangan (2004, 374) similarly described development as a dynamic, complex, and contested concept and draws attention to "the diverse ways in which ideas of development, despite their origins in the West, have been translated, appropriated, refashioned, and reconfigured by local circumstances". Reminiscent of Kapoor's (2008) 'hybridizing strategy', Bebbington (2004, 396) reasoned that "indigenous people may well incorporate the techniques of those who have long been their dominators, and yet do so in a way that strengthens an indigenous agenda pitched in some sense against the interests of those dominating group". Denying the subaltern the opportunity to co-opt and become a part of such hegemony is for Spivak (2010a, 65) a 'primitivist' approach whose implicit effect is to perpetuate subalternity. Questions raised about the Global South's 'right to develop' argument (Norberg-Hodge 2008; Sachs 2002), sometimes due to a presumed "elite control of the framings of problems" (Newell 2005, 72), risk being guilty of such a primitivist and paternalistic approach that implicitly condones Northern consumption and affluence while denying similar aspirations to the emerging economies.

Indeed, in his argument against the Global South's 'right to development', Sachs (2002, 33) claimed that "justice is about changing the rich and not about changing the poor" and the pursuit of development usually only facilitates the integration of the elite of the Global South into the "global circuit of capital and goods" (ibid, 24). Such claims are often responding to the twin fears of ecological collapse reminiscent of the 'limits to growth discourse' of the 1970s (Meadows et al. 1972), as well as of threats posed by growing consumerism in the Global South (Conca 2001), with a tinge of neo-Malthusian anxiety about emerging economies with big populations who are no doubt the racialized 'other' (Ehrlich 1968; Ziser and Sze 2007) and who are subjected to green Orientalism (Lohmann 1993) to rationalize why justice conceived in the form of a Western mode of development is ecologically dangerous and may even be a cultural imposition (Sachs 2002). And while the necessary qualifications are given in such arguments as intended for both North and South (Ehrlich 1968), like population control policies, their impact largely lands on the populations of the Global South. Thus "the ostensive imbalance between responsibility for the damage and obligation for repair" (Castro 1972, 35) of the ecological crisis is further prolonged. This is why critical reflexivity is so important. Even as people in the South may themselves desire a non-industrial decentralized model of development, there is a fundamental contradiction and hypocrisy when a scholar residing in an industrialized country and who lives a life of relative material privilege with immediate consequences for GHG emissions – such as twenty-four-hour access to electricity, refrigerators, private automobiles – prescribes it for the Global South without altering their own way of life first. Even as the Global South is advised to not aspire to lives with these privileges, because they should 'not make the mistakes' the West did, most Western scholars continue to benefit from the very system of industrialized development they might critique or discredit – a disavowal in words and not in deed.

Postcolonial critiques of modernization and development have sometimes led to arguments for alternatives to development and related assertions about authenticity in the kinds of social movements that can challenge the US-based imperial globality and global coloniality (Escobar 2004; Norberg-Hodge 2008). Yet, others caution against a romanticization of such alternatives. Social movements that often get pegged as seeking "alternatives to development", Rangan (2004, 374) argued,

> are in fact very much internal to, and produced by, the processes of uneven geographical development, and articulated within the context of state action. . . . Their mobilization depends on using the legitimizing discourse of development often employed by the states within which they are based, to press for better access to resources or social rights of belonging, or demand that problems of uneven geographical development are addressed.

Clearly there is more than one way in which disenfranchised groups seek justice. Some do so by seeking alternatives, some by working to expand their possibilities within available frameworks, and others may take the available frameworks and seek to transform them. Spivak (2010a, 27–28) cautions against the risk of unnecessarily impeding the project of "counterhegemonic ideological production . . . by making one model of 'concrete experience' the model". Instead of abandoning development altogether, Peet and Hartwick (2015) argued for a critical modernist approach to development that takes seriously feminist, postcolonial, and Marxist critiques of development without abandoning the promise of the idea of development altogether and what it may mean for the dispossessed. According to Simon (2007, 206) "for people near the top of the development pyramid to adopt an antidevelopment stance is politically and/or morally inappropriate if it means abandoning reflexive engagement with poverty". As an interviewee put it poignantly:

> I think those who try and romanticize poverty have a lot of food in their stomach. Then only you can romanticize poverty. And so you can sit in Cambridge and Oxford and romanticize poverty.
> (Interview, CSE official, male, December 2008, Delhi)

Alongside modernization, postcolonial geographers and political ecologists have been fairly unified in their critique of capitalism as obstacles to social justice (McEwan 2003; Nash 2002, 2004; Sachs 2002; Sidaway 2000, 2002; Neumann 2009; Wainwright 2010). While such critique is warranted, I believe we need to pay attention to not only its generalized character – and propensity for injustice and ecological harm – but also to the particular ways in which they have been deployed historically and geographically. Blaut (1993) pointed out, for instance, that industrial capitalism was never the sole product of the West, yet it was European colonization that exacerbated the impacts of a particular form of this phenomenon – resulting in the Industrial Revolution and the consequent rise of capitalism at a global scale – with material consequences for the state of the world today. North/South politics seeks distributive economic and environmental justice within this historically contingent process of the development of capitalism and modernization (Castro 1972; Chakrabarty 2009). Arguments for *replacing* North/South analysis with an analysis of capitalism (e.g. Wainwright 2010) not only seem oblivious to the entanglements of capitalism with coloniality and other hierarchies but reflect what Grosfoguel (2011, 28) has argued is "the pernicious influence of coloniality . . . the first task of a renewed leftist project is to confront the Eurocentric colonialities not only of the right but also of the left". He asserts:

> To call the present world-system "capitalist" is, to say the least, misleading . . . Capitalism is an important constellation of power, but not the sole one. Given its entanglement with other power relations, destroying the capitalist aspects of the world-system would not be enough to destroy the present worldsystem.
>
> (ibid, 13)

Following Grosfoguel (2011, 31), a decolonial intervention in anti–capitalist scholarship involves resisting such economic determinism and anti–modern fundamentalisms to pursue pluriversal approaches to transmodernity, as well as "the eradication of the continuous transfer of wealth from South to North and the institutionalization of the global redistribution and transfer of wealth from North to South".

Figure 3.2 Side event hosted by TERI at COP-15, Bella Center, Copenhagen, December 2009

India's postcolonial politics and the backlash

India embodies a dual identity of a formerly colonized climate vulnerable Third World country and a top emitting emerging economy. In climate negotiations, India has clearly pushed for a Global South agenda of seeking justice from the Global North. North/South politics, first institutionalized in the NIEO proposal by postcolonial states, was about bringing the process of decolonization to its logical completion by challenging the hegemony of the Global North in a postcolonial world (Bhagwati 1977; Cox 1979; Doty 1996). Gilman (2015, 1) has asserted that:

> the NIEO was the most widely discussed transnational governance reform initiative of the 1970s. Its fundamental objective was to transform the governance of the global economy to redirect more of the benefits of transnational integration toward "the developing nations" – thus completing the geopolitical process of decolonization and creating a democratic global order of truly sovereign states. It was, in short, a proposal for a radically different future than the one we actually inhabit.

Although the proposal never came into fruition owing to its rejection by the United States, the desire for subverting the status quo has clearly persisted. Global environmental change provided the opportunity for a reemergence of the discourse of North/South justice, seemingly giving the South more bargaining power. India's participation in the climate negotiations needs to be looked at in this historical context. In labeling India's climate politics a postcolonial politics, I am referring to a number of representational practices that have material ramifications for the decolonial project. Firstly, in insisting on India's Global South identity juxtaposed against an environmentally irresponsible North, the Indian officials engaged in a binary 'othering' process that have typically been understood to be the domain of dominant groups in colonialism. While postcolonial critique has tended to focus on the North's discursive representation of an essentialized identity of the South, my research revealed that when the colonial gaze is returned, Southern actors also engage in similar representational practices (Doty 1996). The use of binary categories – developed and developing – is a strategic one for Indian negotiators from the point of view of avoiding mandatory mitigation commitments, and the economic costs that go with them, that are deemed to be the responsibility of developed countries. Therefore binary othering is not only a colonial strategy but a counter-colonial one as well. Postcolonial scholars should therefore pay attention to who uses binary othering as a strategy and towards what ends, rather than rejecting this representational practice a priori due to its colonial associations.

Secondly, such representational practices are intensified by 'organic crises' that threaten to destabilize hegemony (Doty 1996, 8). The predicament of global environmental crisis has arguably served as such an organic crisis that threatens to challenge the status quo of the world order. With greater bargaining power accruing to the Global South due to its crucial role in the international cooperation required to combat global environmental crises (Roberts and Parks 2007), Indian negotiators strove to insert clear provisions for North/South equity in the climate convention emerging in the late 1990s. Pitted against the US agenda of inserting the notion of 'common responsibility' – a discourse of one-worldism (Agarwal and Narain 1990) – India wanted to emphasize differentiated responsibilities. The subsequent CBDRRC was a compromise, open to interpretation, but the insertion of the Annex I/non-Annex I distinction in the Kyoto Protocol was a clear victory. It enabled a particular interpretation of differentiated responsibility as an equity framework and its "institutional validation" (Spivak 2010b, 228). Indian negotiators and intellectuals worked hard to preserve this binary distinction during the so-called post-Kyoto discussions and in so doing were defending a geopolitical spatial/scalar imaginary that had been institutionalized in a global treaty after a hard-won battle. As then Intergovernmental Panel on Climate Change (IPCC) Chair and Director-General of TERI, Dr. Rajendra Pachauri, said, "Nobody is trying to shirk responsibilities, but we shouldn't try to water down something that has been achieved after long negotiations and battles" (Participant observation, COP-15, Bella Center, Copenhagen, December 2009). The Indian position has been predicated on problematizing entrenched differentials in power, material privileges, and GHG emissions between the average citizen in more and less powerful states by asserting the "strategic essentialism" (Spivak 1993) of the Global South imaginary. Viewing "essentialist notions of identity" as strategic social constructs that can be made by hegemonic as well as marginalized groups (Jacobs 1996, 162) helps us understand India's agenda in climate negotiations as a decolonial hybridizing strategy. Since "theoretical generalizations about the socially constructed nature of essentialized identity have an uneven political consequence which is far from incidental to ongoing political struggles" (Jacobs 1996, 163), the scholarly dismissal of a North/South politics of climate change is disturbingly complicit with the United States negotiating strategy in climate discussions. A Western scholar's disavowal of the North/South frame of binary difference as lacking validity is a privileged point of view, enabled by the conflation of a Eurocentric panopticon (Toal 1994) with universality. As Said (1994, 48) suggested, such scholarly privilege fits uncomfortably well with a deliberate imperial agenda. Interestingly, even as binary constructions are critiqued, scholars themselves are unable to escape binary ways of thinking – such as between subaltern and elite – a phenomenon I call the paradox of binaries, after Jones' (2009) paradox of categories.

A third aspect of a postcolonial representational politics is that hegemonic groups adopt new strategies to reinscribe their power, confirming the adaptive and arbitrary nature of (post)colonial identity construction (Doty 1996; Jacobs 1996; Kapoor 2008). In climate negotiations, this is shown by the US negotiating team's tireless and somewhat successful attempts to establish a binary distinction within the category of 'developing countries' – 'emerging economies' and the rest. But the "hype created about the Indian middle class . . . is not a new politics", one of my interviews shared with me, pointing out that similar arguments were made at Rio Summit negotiations (Interview, CSE official, male, December 2009, Bella Center, Copenhagen). And while the US strategy may seem as an effort to transcend colonial binaries (of West and Third World), colonial attitudes persist. As Doty (1996) and Sidaway (2002) have pointed out, the term 'emerging' portrays the non-Western world both as a threat and requiring paternalistic intervention. This is apparent in the emphasis on MRV that have accompanied any commitment to climate finance made by Annex I countries and that was in turn perceived as a tool of neocolonialism by my interviewees. Even scholars who acknowledge the need for a flow of resources from North to South towards equalizing growth are not immune to such paternalism, warning that any funds given for adaptation or mitigation should be "carefully managed . . . to ensure that these benefits are actually realized" (Roberts and Parks 2007, 155). Many of my interview respondents, not too surprisingly, bristled at the righteousness of the paternalistic Western academic or environmentalist perspectives that chastise India for its coal-fired power plants, its internal inequity issues, or seek to place it in a different category due to its growing economy. Such perspectives are seen as neocolonial and compatible with the US negotiation position that was perceived to be 'hiding behind the emerging economies' in attempting to deflect attention away from its own obligations. This apparent complicity is troubling because it implicitly condones Northern consumption and energy use while painting China and India as the problem, a case of Orientalism in the guise of neo-Malthusian environmental concern (Ziser and Sze 2007). This particular strategy seeks to silence both 'elites' and 'emerging economies' by demonizing the seemingly burgeoning middle-class Chinese and Indian consumers and their ravenous appetites that threaten climate collapse. Jacobs (1996, 2–3) attributes such 'racialized othering' to the need for colonial powers to "anxiously [reinscribe]" the binaries of "a 'demonized Other' versus legitimate 'metropolitan Self' . . . in the face of their contested or uncontainable certainty". Coupled with such blatant Orientalism and the more benign-seeming green Orientalism (Lohmann 1993) that associates the Global South with lower impact ecologically sustainable lifestyles, the Global North can have its cake and eat it too – denying North/South equity while perpetuating colonial stereotypes in one fell swoop.

We may never know if scholarly critique of a North/South binary and its alignment with the US negotiating team's strategy of insisting on differentiation of non-Annex I were more than coincidental. In any case, we are at a juncture when the Kyoto Protocol – which institutionalized the North/South binary and assigned differential responsibilities accordingly for addressing the climate crisis – is bowing out to make way for an Agreement that does not recognize differential responsibilities, before any serious attempts were even made to address those outlined in the Kyoto Protocol. This is a good moment (as any other) to reassess the relationship between critique and complicity. In analyzing my Delhi and Copenhagen interviews for my dissertation, I found it difficult to dismiss North/South climate justice claims as "ideological posturing by elites" (Berger 2004). Further, the championing of a North/South framing of the climate debates was not done simply by those I would consider the Indian elite, that is, at seniormost positions of authority. My respondents encompassed a range of officials – Indian representatives for UNFCCC negotiations, other bureaucrats holding offices in government and nongovernment institutions, civil society members including officials – junior to senior level – of domestic as well as international environmental nonprofits, news reporters, professors. They were unanimous in asserting India's identity as a developing country with competing priorities that rendered India unable to accept binding emission limits in the foreseeable future, even as some of them spoke to the need for accountability and consistence across scales. Berger's (2004) notion of 'Third Worldism' problematically assumes that the Third World elite that use this rhetoric do so to further their class interests and that they do not speak for the subaltern – in this case the most climate-vulnerable citizens in India. This may or may not be true, but to assume so might effectively contribute to silencing the subaltern. Following Spivak (2010a), if the 'authentic' subaltern cannot speak in a way that can be heard or understood by hegemonic groups, and the so-called Third World elite who seek to be their voice cannot be trusted to be authentic enough, then we have quite the predicament. It then inevitably falls on the First World intellectual to rescue the Third World subaltern from their elite. This paradox can be extrapolated to the silencing of 'emerging economies' speaking for the Global South as well (Joshi 2015). This is problematic especially when we (First World academics) recognize that these assumptions and thought patterns can lead to the legitimizing of the sort of international politics that have given us the Paris Agreement, even as we continue to benefit from the North/South status quo, and even as our carbon footprint per capita far exceeds that in the Global South.

The ecological debt of colonialism

North South environmental justice claims have often been expressed as an ecological debt (Martinez-Alier 2002). This is an idea that is often attributed to Southern NGOs, including an Ecuadorian NGO Accion Ecologica that defined ecological debt in 1999 as "[t]he debt accumulated by Northern, industrialized countries toward Third World countries on account of resource plundering, environmental damages, and the free occupation of environmental space to deposit wastes, such as greenhouse gases, from the industrial countries" (cited in Rice 2009, 227). As such, the notion of ecological debt historicizes contemporary patterns of asymmetric power relations and unequal appropriation of the global commons by placing them in the context of colonialism: "It is well documented how, through the colonial period and the industrial revolution up till now, natural resources have been flowing from South to North" (Goeminne and Paredis 2010, 698). This is not just a rhetorical claim based on ideology or anecdotal evidence (Rice 2009) but one documented through rigorous research (Parks and Roberts 2010; Srinivasan et al. 2008). Through a lens of ecological debt, the American lifestyle that George H. W. Bush was unwilling to negotiate in 1992 is understood to have been made possible by patterns of ecologically unequal exchange, ecological aggression, and unaccounted-for socio-ecological subsidies that have enabled the Global North to live beyond their ecological means (Bosselmann 2004; Martinez-Alier 2002; Rice 2009). As such there is a planetary ecological debt the North owes to the South, one overlooked in international relations and climate negotiations (Goeminne and Paredis 2010). The idea of ecological debt therefore is a call for justice in the form of reparations, owed by countries – and companies – that benefited mostly from colonialism and continue to benefit from neo-colonialism. The demands for reparations are due to past and continuing disproportionate encroachment on environmental space without payment and without recognition of other countries' entitlements to that space (Martinez-Alier 2002).

The notion of ecological debt has been utilized by some scholars to advocate for historical responsibility to shape a fair burden-sharing agreement for climate mitigation. This entails

> the acceptance of accountability for the full consequences of industrialization that relied on fossil fuels. . . . The contention of the proponents of the application of historical responsibility to climate change is that the nations that have controlled the process of industrialization, and have benefitted the most from industrialization, should restore the playing field to a level position by bearing most of the costs that are resulting from the accumulated greenhouse gases injected into the atmosphere by industrialization.
>
> (Shue 2009, 1)

Arguments for financial and technological transfers have been a mainstay of North/South negotiations and are present in many INDCs submitted by Global South countries. Southern governments have, however, been reluctant to explicitly push the concept of ecological debt during negotiations for fear of alienating Northern governments (Martinez-Alier 2002). This sentiment was confirmed by my interviews. While acknowledging that the language of 'ecological debt' is not explicitly used in climate negotiations, the idea was appealing to my interviewees because it disrupts the idea of financial and technological transfers from Annex I to non-Annex I countries as charity or aid, as it is typically portrayed. During COP-15 in 2009, one of the side events hosted by TERI at the Bella Center presented a proposal for quantification of climate debt liabilities of select Global North countries. The Indian government has likewise articulated that any funds to be received by developing countries, including India, for mitigation and adaptation of climate change should be seen as "entitlement not aid" (GoI 2009, 41). The INDC submitted by the country does not mention this explicitly but does mention that developed countries have a historical responsibility to mitigate climate change.

The acceptance of the idea of climate debt and reparations would transform the way in which resource transfers are discussed in climate financing, given the predominance of paternalistic interventions and conditionalities that typically accompany aid. Doty's (1996, 128) pronouncement of how "foreign aid . . . made possible new techniques within an overall economy of power in North/ South relations . . . [putting] in place permanent mechanisms by which the 'third world' could be monitored, classified, and placed under continual surveillance" is relevant today in Annex I countries' insistence that MRV mechanisms accompany any adaptation funds. Goeminne and Paredis (2010, 693) suggest that "ecological debt is a way of looking at North/South relations and sustainability issues from a Southern peoples' point of view. . . . [It] draws attention to how the present situation has grown out of the often violent and unjust past. It points to the collective responsibility of industrialized countries and companies in relation to socio-ecological problems". The idea of ecological debt is a call for justice that has emerged from civil society in the Global South, and it has been supported by empirical research and critical theory, yet has typically been dismissed as 'rhetoric' by Western negotiators and in Western media as "distracting and unconstructive" (Parks and Roberts 2010, 141; Vihma 2011). It has also been critiqued for being vague on the details – how to quantify the debt, and identify who will pay what to whom and how to prevent elite capture, and so on – as well as for not being sufficiently informed by a genuine understanding of vulnerability and adaptation on the ground, even as it is recognized as a 'populist' political ecology discourse that "draws its authority both from science and from a moral imperative" (Adger et al. 2001, 699). The notion of cultural parallax as it applies to the North/South spatial imaginaries may possibly apply to this notion as well (Nabhan 1998). As Goeminne and Paredis (2010, 698) aptly put it, "for most inhabitants of industrialized countries, it is a most unusual and uneasy way of looking at their own position".

Challenges of quantification and operationalization of debt repayment (or cancellation) aside, it is also the case that there are two other dimensions of the notion of ecological debt. Beyond the more straightforward notion of international distributive justice – difficult as it may be to operationalize – the ecological debt discourse is also about reparative justice and structural change. Describing the notion of climate reparations in terms of moral obligations particularly towards the climate vulnerable, Burkett (2009) has emphasized the need for acknowledgment of harm done – an apology – before work can be done to repair harmful patterns. Secondly, acknowledging an ecological debt also entails a commitment to pursue development differently, eschewing colonial patterns of exploitation that impoverished land, culture, and people in the Global South in the name of development that only benefits a select few elites (Rice 2009). In this sense ecological debt claims seek to exert a counterhegemonic discourse about development, resource use, and well-being, from the perspective of the subaltern that may well be critical of the development agendas of Global South governments. With the exception of the Bolivarian Alliance, not many G-77 members have taken a stand during climate negotiations against neoliberal capitalist modes of development that are hegemonic in the world today. Similarly, the Indian government's approach to development was a point of consternation for a small number of individuals I had interviewed in India, even as they vociferously defended India's position on the historical responsibility of the Global North for climate change. An academic and environmental activist shared this critique:

> The government is, essentially, aping itself on big capitalist governments. And it's also arming itself with all that it thinks it needs to be a great power. So it's spending billions of dollars, in the next couple of years, probably five billion dollars, on arms. . . . So these are not signs of a country which wants to go into sustainable growth at all! India has a very old tradition as a leader of the South, the Third World, the Non-Aligned Movement. So, our diplomats and all are very clever, but, there is a mismatch between what they say they're going to do and what they actually do.
>
> (Interview, JNU Professor, male, December 2008, Delhi)

India's postcolonial climate politics: a counter-hegemony in the making?

A Gramscian approach to understanding hegemony is useful in analyzing India's postcolonial climate politics. The persistence of a North/South struggle in international economic and environmental negotiations is a reminder that the project of decolonization is far from over. Even as many formerly colonized countries have achieved juridical-political decolonization at the level of nation-states, a condition of coloniality of power remains in core-periphery relations that continue within a paradigm of patriarchal-modernist-capitalist-colonialism (Grosfoguel 2011). For Gramsci, a successful counterhegemonic movement is able to supplant a prevailing hegemonic order when counterhegemonic ideas are sufficiently developed and accepted by a polity at the level of socio-cultural, economic, and political structures and institutions, through the development of a 'historic bloc' that can withstand co-optation efforts by hegemonic groups – what Gramsci refered to as "*trasformismo*" (Femia 1987) and Omi and Winant (2015, 173) refer to as "entrism". Viewing core-periphery relations and contemporary climate politics through this lens, it would seem that for the project of decolonization to be complete through a successful counterhegemonic movement, leaders in the Global South would need to not only oppose dominance by the Global North and seek more power within the power structure but challenge colonial ideologies and practices, as in the more expansive notion of ecological debt, so that a hegemonic culture of dominance is replaced by a more egalitarian and sustainable way of being. Further, efforts to thwart hegemony at the international scale may not succeed when they are not supported by groundwork laid in the domestic sphere: "the task of changing world order begins with the long, laborious effort to build new historic blocs within national boundaries" (Cox 1993, 65). It is possible that an absence of co-ordination between power relations within the Global South and the sphere of North/South politics has played a role in rendering unsuccessful so far the North/South struggle in the context of NIEO and again in the Kyoto Protocol. Meanwhile, the connection between a North/South decolonial politics and that of struggles within the Global North is worthy of investigation as well. Ongoing struggles for racial justice and Indigenous sovereignty within the US context offer interesting parallels to struggles for North/South equity.

Mainstream Indian climate politics does not yet constitute a counter-hegemony to the prevailing world order in the Gramscian sense of term, as it does not seek to circumvent the hegemony of a capitalist mode of growth and development ruled by the Global North. India's 'right to development' is largely focused on economic growth and industrialization, often disregarding their consequences for environmental sustainability and justice for disenfranchised populations within (Williams and Mawdsley 2006). India's climate politics can best be described as a resistance to neocolonial *dominance* by hegemonic states, particularly the United States. For India's climate politics to be counterhegemonic in the Gramscian sense, the Indian state would have to eschew its authoritarian, technocratic, and centralized approach to development. Even as the state pursues a capitalist and modernist development agenda, environmentalists and social justice advocates have long worked to push back against the state's agenda without renouncing the notion of development altogether (Agarwal and Narain 1990; Rangan 2004; Roy 2014; Shiva 2008). NGOs such as Navdanya and CSE question India's pursuit of economic growth in the name of pro-poor development and instead advocate for more equitable, decentralized, and earth-centric forms of development. There were clear signs of discontent shared during interviews about India's industrialization- and economic growth-centered development. Some respondents explicitly articulated a Gandhian approach to development – or *swaraj*, emphasizing autonomous development based on self-governance and designed to meet its citizens' needs and aspirations – as opposed to a consumptive approach leading to 'luxury' emissions. From a Gramscian perspective, these ideas would need to be scaled up and become hegemonic within the country if they are to replace hegemonic ideas of progress. The official positions of the states of Maldives and Bolivia at COP-15 and COP-16 respectively, corresponding with the "People's Agreement of Cochabamba" at the World People's Conference on Climate Change and the Rights of Mother Earth on April 24, 2010, hint at the possibility of such movements being scaled up and entering international discourse. The championing of a Gandhian approach to development in India may indicate the possibility of popular support for such a movement to be possible in India as well. India's INDC pays homage to Gandhi and a commitment to "Development without Destruction" but for the most part espouses a techno-centric and modernist development modality.

The INDC claims that the Constitution of India supports ecologically sensitive development: "The State shall endeavour to protect and improve the environment and to safeguard the forests and wildlife of the country" (India n.d., 7). But as contestations in 2019 over India's proposed amendments to the Forest Act, the Citizenship Amendment Act, the move to strip Kashmir of constitutional autonomy, and the Supreme Court's ruling on Ayodhya all demonstrate, the state is often at odds with the interests of its citizens, many of whom are actively disenfranchised from their de facto citizenship in a state leaning towards Hindu nationalism and settler colonialism (Kukreti 2019; Shankar 2020; Wani 2020). From this perspective the notion of a state's sovereignty in international negotiations needs unpacking, when there are competing sovereignty claims from the "high-modern extractive nation-state" and those with "Indigenist, ethno-regionalist" claims to nationalism (Sivaramakrishnan and Cederlof 2006). A North/South politics that is detached from these undercurrents appears unlikely to be successful in destabilizing colonial logics. On the one hand, there is the question of legitimacy arising from consistency of justice claims at multiple scales: does a North/South justice claim made at the site of international climate negotiations beholden the Indian state to take seriously similar justice claims against its neocolonial and settler colonial tendencies? I believe it does. On the other hand, the absence of a robust domestic counterhegemonic politics precludes a successful counterhegemonic movement in the international sphere (Cox 1993). For a nation-state whose climate agenda has been to hold hegemonic powers accountable to past and continuing harm, it is important to remain accountable to those over whom one is hegemonic. In this sense, not only has the Indian state's record on authoritarianism against its Indigenous, Muslim, and other marginalized populations been consistently poor; its bilateral relations with neighbors in South Asia have been problematic as well (Zehra 2019; Bose 2020). These dynamics are in many ways beset by the vestiges of British colonialism (Bose and Jalal 2011), and India's approach in each of these struggles needs to be held accountable to its North/South justice claims. I believe the Indian state's legitimacy in its struggles for equity with the Global North hinges on its responses to domestic and regional counterhegemonic struggles.

The climate crisis is an "organic crisis" that enables the conditions for counter-hegemony (Cox 1993; Doty 1996). A true counter-hegemony adopts 'third space' as a strategy rather than a "hybridizing strategy" (Kapoor 2008) – reminiscent of the Third World Project (Prashad 2014) – and is more cognizant of the ecological challenges posed by the climate crisis, than the Indian government's resolute pursuit of large-scale industrialization that may even be inimical to the interests of the country's poor and marginalized on whose behalf it claims the right to development and atmospheric space. Yet India's efforts to resist US dominance in global economic and environmental governance, and to seek to hold it accountable for its exploitation of the earth's resources, has been commendable in the context of climate debates. Kapoor (2008) suggests that the postcolonial strategies of 'third space' and 'hybridization' need not be mutually exclusive and can be pursued in tandem. Gramsci himself seemed to allow for "an infinite range of strategies" of resistance to be possible (Femia 1987, 55). Equally important is to heed what Gramsci called '*trasformismo*' (Cox

1993; Femia 1987) – the potential of cooptation of leaders of subaltern groups by hegemonic powers, thereby weakening the formation of a historic bloc and eventually a successful counter-hegemony – prompting Kapoor (2008, 144) to acknowledge that a hybridizing strategy "can be put to regressive as much as progressive use". One of the biggest dangers of such *trasformismo* is that a leading member of a counterhegemonic struggle may betray their constituents once the gains of a political struggle are realized. While it is important to always be alert to this potential – and to empower subaltern groups to hold their representatives in struggle more accountable – assuming its existence a priori is problematic. Similarly, accusing Indian intellectuals of *trasformismo* simply because of their articulation of modernization-as-development is problematic. Some of my respondents were clearly vague and ambivalent in their articulation of India's 'right to development' – seeing it as necessary but not sufficient for climate justice – fluctuating between ecological concerns and material concerns for the poorest. For Kapoor (2008), such ambivalence is characteristic of hybrid anti-colonial subjectivities. To what extent resistance to dominance by hegemonic states that uses 'hybridization' as a strategy paves the way for a genuine counter-hegemony against the dominant system that uses 'third space' as a strategy is an unanswered question at this time.

While I have pointed to the limits of India's postcolonial politics amid an absence of a 'historic bloc' within India and within the Global South for a more successful North/South counterhegemonic politics, it is possible that there is a role here also for counterhegemonic politics within the Global North. Indian environmentalists and critics of the Indian state have long pointed to the injustice of the Global North's demands for ecologically benign development for India while ignoring the need for a similar paradigm shift in the Global North (Agarwal and Narain 1990). Others have similarly pointed to the limits of decolonization efforts constrained by national boundaries in a condition of global coloniality (Grosfoguel 2011; Solon 2019). If the success of a counter-hegemonic movement in international relations is predicated on the creation of a 'historic bloc' in the domestic sphere (Cox 1993), what might it mean for the role of counterhegemonic movements in the Global North for the success of North/South justice efforts? While calls for alliance-building between social movements in the North and South have been made often (e.g. Pellow 2007), conceptual connections may need to be made to identify the power-hoarding strategies of a hegemon. I will focus on a few points of connection here on the positions taken by the United States against justice claims arising from multiple Global South locations. Omi and Winant's (2015, 109) analysis of the paradigm of color-blindness as the "hegemonic concept of race in the US", one that seeks to maintain White supremacy while refusing to see race, has a similar effect as the argument for one-worldism (Agarwal and Narain 1990), which seeks to obstruct structural transformations for equity by invalidating the presence of North/South inequities. Similarly, there are parallels between US disrespect of sovereign nations abroad as well as within the United States (Dunbar-Ortiz 2014). Its reluctance to accept international environmental treaties with teeth could be seen as a desire to avoid the kinds of legal obligations it is already committed to within its borders, and which it tries to avoid whenever possible.

It is important for US racial hegemony to be understood as not only limited to the borders of the United States but one that is extended to Third World countries as well (Ziser and Sze 2007). Challenging White supremacy within the borders of the United States by calling on the US constitution's commitment to equality among its people and by calling on the United States to be accountable to treaties signed with Indigenous Nations is important for climate justice; but given the United States' historical responsibility for climate change and ecological debts owed to the Global South, these critical perspectives need to be broadened to a global context where US influence is exercised through imperialism and neoliberal capitalism. For people who live and work in the United States, US-centrism operates invisibly – it is a kind of state-centrism that does not get named as such. Likewise, efforts to minimize North/South justice claims as well as racial/Indigenous justice claims are sometimes countered with an invitation to center a class-centric analysis (Lichtenstein 2005; Omi and Winant 2015), which is also an exercise in Eurocentrism and White supremacy (Grosfoguel 2011). Working against these tendencies calls for a common strategy that is

> anti-capitalist, anti-patriarchal, anti-imperialist and against the coloniality of power towards a world where power is socialized, but open to a diversality of institutional forms of socialization of power depending on the different decolonial epistemic/ethical responses of subaltern groups in diverse locations of the world-system.
>
> (ibid, 32)

A successful North/South counterhegemonic movement appears to call for a scaling down of North/South climate justice claims and horizontal collaborations within the Global South as well as a scaling up of domestic climate justice politics in countries such as the United States. A recognition of disparate struggles as valid and a recognition of the multifarious strategies of maintaining hegemony and the logics of coloniality are critically needed.

References

Adger, W. Neil, Tor A. Benjaminsen, Katrina Brown, and Hanne Svarstad. 2001. "Advancing a Political Ecology of Global Environmental Discourses." *Development and Change* 32: 681–715.

Agarwal, Arun, and Sunita Narain. 1990. *Global Warming in an Unequal World*. New Delhi: Centre for Science and Environment.

Anand, Ruchi. 2004. *International Environmental Justice*. Burlington: Ashgate.

Barnett, Jon. 2007. "The Geopolitics of Climate Change." *Geography Compass* 1, no. 6: 1361–75.

Bebbington, Anthony. 2004. "Movements and Modernizations, Markets and Municipalities." In *Liberation Ecologies*, edited by Richard Peet and Michael Watts, 2nd ed. London and New York: Routledge.

Berger, Mark T. 2004. "After the Third World? History, Destiny and the Fate of Third Worldism." *Third World Quarterly* 25, no. 1: 9–39.

Bhabha, Homi. 1994. *The Location of Culture*. New York: Routledge.

Bhagwati, Jagdish N., ed. 1977. *The New International Economic Order*. Cambridge, MA: The MIT Press.

Blaikie, Piers, and Harold Brookfield. 1987. *Land Degradation and Society*. London: Methuen & Co. Ltd.

Blaut, James M. 1993. *The Colonizer's Model of the World*. New York: The Guilford Press.

Blunt, Alison, and Cheryl McEwan. 2002. "Introduction." In *Postcolonial Geographies*, edited by Alison Blunt and Cheryl McEwan. New York: Continuum.

Bose, Sugata, and Ayesha Jalal. 2011. *Modern South Asia*, 3rd ed. New York: Routledge.

Bose, Tapan K. 2020. "The Kalapani Ibroglio: Has India Pushed Nepal too Far?" *The Wire*, May 26.

Bosselmann, Klaus. 2004. "In Search of Global Law: The Significance of the Earth Charter." *Worldviews: Environment Culture Religion* 8, no. 1: 62–75.

Bryant, Raymond L. 1998. "Power, Knowledge and Political Ecology in the Third World: A Review." *Progress in Physical Geography* 22, no. 1: 79–94.

Bryant, Raymond L., and Lucy Jarosz. 2004. "Thinking About Ethics in Political Ecology." *Political Geography* 23: 807–12.

Burkett, Maxine. 2009. "Climate Reparations." *Melbourne Journal of International Law* 10, no. 2.

Castro, João A. D. A. 1972. "Environment and Development: The Case of Developing Countries." In *Green Planet Blues, 2010*, edited by Ken Conca and Geoffrey D. Dabelko. Boulder: Westview Press.

Chakrabarty, Dipesh. 2009. "The Climate of History: Four Theses." *Critical Inquiry* 35, no. 2: 197–222.

Conca, Ken. 2001. "Consumption and Environment in a Global Economy." *Global Environmental Politics* 1, no. 3: 53–71.

Cowan, Sam. 2015. "The Indian Checkposts, Lipu Lekh, and Kalapani." *The Record*, December 14.

Cox, Robert W. 1979. "Ideologies and the NIEO: Reflections on Some Recent Literature." *International Organization* 33, no. 2: 257–302.

———. 1993. "Gramsci, Hegemony and International Relations: An Essay in Method." In *Gramsci, Historical Materialism and International Relations*, edited by Stephen Gill. Cambridge: Cambridge University Press.

Damodaran, Vinita. 2006. "Indigenous Forests." In *Ecological Nationalisms*, edited by Gunnel Cederlof and K. Sivaramakrishnan. Seattle: University of Washington Press.

Doty, Roxanne L. 1996. *Imperial Encounters*. Minneapolis: University of Minnesota Press.

Dunbar-Ortiz, Roxanne. 2014. *An Indigenous Peoples' History of the United States*. Malden, MA: Beacon Press.

Ehrlich, Paul R. 1968. *The Population Bomb*. New York: Ballantine Books.

Ekers, Michael, Alex Loftus, and Geoff Mann. 2009. "Gramsci Lives!" *Geoforum* 40: 287–91.

Escobar, Arturo. 2004. "Beyond the Third World: Imperial Globality, Global Coloniality and Anti-Globalisation Social Movements." *Third World Quarterly* 25, no. 1: 207–30.

Femia, Joseph V. 1987. *Gramsci's Political Thought*. Oxford: Clarendon Press.

Gandhi, Leela. 1998. *Postcolonial Theory: A Critical Introduction*. New York: Columbia University Press.

Gilman, Nils. 2015. "The New International Economic Order: A Reintroduction." *Humanity* 6, no. 1: 1–16.

Goeminne, Gert, and Erik Paredis. 2010. "The Concept of Ecological Debt: Some Steps Towards an Enriched Sustainability Paradigm." *Environment, Development, Sustainability* 12: 691–712.

Government of India. 2009. *Climate Change Negotiations.* New Delhi: Ministry of Environment and Forests.

Gramsci, Antonio. 1957. *The Modern Prince & and Other Writings.* Translated by Louis F. Marks. New York: International Publishers.

Grosfoguel, Ramon. 2011. "Decolonizing Post-Colonial Studies and Paradigms of Political-Economy: Transmodernity, Decolonial Thinking, and Global Coloniality." *Transmodernity: Journal of Peripheral Cultural Production of the Luso-Hispanic World* 1, no. 1: 1–38.

India. n.d. "INDC: India's Intended Nationally Determined Contribution: Working Towards Climate Justice." www.indiaenvironmentportal.org.in/content/419700/indias-intended-nationally-determined-contribution-working-towards-climate-justice/.

Jacobs, Jane M. 1996. *Edge of Empire: Postcolonialism and the City.* New York: Routledge.

Jarosz, Lucy. 2004. "Political Ecology as Ethical Practice." *Political Geography* 23: 917–27.

Jones, Reece. 2009. "Categories, Borders and Boundaries." *Progress in Human Geography* 33, no. 2: 174–89.

Joshi, Shangrila. 2015. "Postcoloniality and the North/South binary Revisited: The Case of India's Climate Politics." In *The International Handbook of Political Ecology*, edited by Raymond L. Bryant. Cheltenham: Edward Elgar.

Kapoor, Ilan. 2008. *The Postcolonial Politics of Development.* New York: Routledge.

Kim, Soyeun. 2009. "Translating Sustainable Development: The Greening of Japan's Bilateral International Cooperation." *Global Environmental Politics* 9, no. 2: 24–51.

Kukreti, Ishan. 2019. "Government Withdraws Proposed Changes to Indian Forest Act." In *Down to Earth.* New Delhi: Center for Science and Environment, November 15.

Laclau, Ernesto. 1990. *New Reflections on the Revolution of Our Time.* New York: Verso.

Lemanski, Charlotte, and Stéphanie T. Lama-Rewal. 2013. "The 'Missing Middle': Class and Urban Governance in Delhi's Unauthorised Colonies." *Transactions of the Institute of British Geographers* 38, no. 1: 91–105.

Lichtenstein, Nelson. 2005. "Trashing Identity Politics: Does It Really Get Us Back to Class?" *International Labor and Working-Class History* 67: 42–49.

Lohmann, Larry. 1993. "Green Orientalism." *Ecologist* 23, no. 6: 202–4.

Manandhar, Mangal S., and Hriday Koirala. 2001. "Nepal-India Boundary Issue: River Kali as International Boundary." *Tribhuvan University Journal* 23, no. 1: 1–21.

Mann, Geoff. 2009. "Should Political Ecology Be Marxist? A Case for Gramsci's Historical Materialism." *Geoforum* 40: 335–44.

Martinez-Alier, Joan. 2002. "Green Justice – Ecological Debt and Property Rights on Carbon Sinks and Reservoirs." *Capitalism Nature Socialism* 13, no. 1: 115–19.

McEwan, Cheryl. 2003. "Material Geographies and Postcolonialism." *Singapore Journal of Tropical Geography* 24, no. 3: 340–55.

Meadows, Donella H., Dennis L. Meadows, Jorden Randers, and William W. Behrens. 1972. *The Limits to Growth.* New York: Universe Books.

Nabhan, Gary P. 1998. "Cultural Parallax in Viewing North American Habitats." In *Environmental Ethics*, edited by Richard Botzler and Susan Armstrong, 2nd ed. New York: McGraw Hill.

Nash, Catherine. 2002. "Cultural Geography: Postcolonial Cultural Geographies." *Progress in Human Geography* 26, no. 2: 219–30.

———. 2004. "Postcolonial Geographies: Spatial Narratives of Inequality and Interconnection." In *Envisioning Human Geographies*, edited by Paul Cloke, Philip Crang, and Mark Goodwin. London: Arnold.

Neumann, Roderick P. 2009. "Political Ecology: Theorizing Scale." *Progress in Human Geography* 33, no. 3: 398–406.

Newell, Peter. 2005. "Race, Class and the Global Politics of Environmental Inequality." *Global Environmental Politics* 5, no. 3: 70–94.

Norberg-Hodge, Helena. 2008. "The North/South Divide." *Ecologist* 38, no. 2: 14–15.

Omi, Michael, and Howard Winant. 2015. *Racial Formation in the United States*, 3rd ed. New York: Routledge.

Parenti, Christian. 2011. *Tropic of Chaos*. New York: Nation Books.

Parks, Bradley C., and J. Timmons Roberts. 2010. "Climate Change, Social Theory and Justice." *Theory, Culture and Society* 27: 134–66.

Peet, Richard, and Elaine Hartwick. 2015. *Theories of Development*, 3rd ed. New York: The Guilford Press.

Pellow, David N. 2007. *Resisting Global Toxics*. Cambridge, MA: The MIT Press.

People's Agreement of Cochabamba. 2010. "World People's Conference on Climate Change and the Rights of Mother Earth." April 24. https://pwccc.wordpress.com/2010/04/24/peoples-agreement.

Power, Marcus. 2006. "Anti-Racism, Deconstruction and 'Overdevelopment'." *Progress in Development Studies* 6, no. 1: 24–39.

Prashad, Vijay. 2014. *The Poorer Nations*. New York: Verso.

Rangan, Haripriya. 2004. "From Chipko to Uttaranchal: The Environment of Protest and Development in the Indian Himalaya." In *Liberation Ecologies*, edited by Richard Peet and Michael Watts, 2nd ed. New York: Routledge.

Rice, James. 2009. "North/South Relations and the Ecological Debt: Asserting a Counter-Hegemonic Discourse." *Critical Sociology* 35, no. 2: 225–52.

Robbins, Paul. 2011. *Political Ecology*, 2nd ed. West Sussex: Blackwell.

Robbins, Paul, and Julie Sharp. 2003. "The Lawn-Chemical Economy and Its Discontents." *Anitpode* 35, no. 5: 955–79.

Roberts, J. Timmons, and B. C. Parks. 2007. *A Climate of Injustice*. Cambridge, MA: The MIT Press.

Rothermund, Dietmar. 2008. *India*. New Haven: Yale University Press.

Roy, Arundhati. 2004. *An Ordinary Person's Guide to Empire*. Cambridge: South End Press.

———. 2014. *Capitalism*. Chicago: Haymarket Books.

Sachs, Wolfgang. 2002. "Ecology, Justice, and the End of Development." In *Environmental Justice*, edited by John Byrne, Leigh Glover, and Cecilia Martinez. New Brunswick: Transaction Publishers.

Said, Edward W. 1978. *Orientalism*. New York: Pantheon Books.

———. 1994. *Culture and Imperialism*. New York: Vintage Books.

Shankar, Saumya. 2020. "India's Citizenship Law, in Tandem with National Registry, Could Make BJP's Discriminatory Targeting of Muslim's Easier." *The Intercept*, January 30.

Shiva, Vandana. 2008. *Soil Not Oil*. New York: South End Press.

Shue, Henry. 2009. "Historical Responsibility." SBSTA Technical Briefing, June 4. https://unfccc.int/files/meetings/ad_hoc_working_groups/lca/application/pdf/1_shue_rev.pdf.

Sidaway, James. 2000. "Postcolonial Geographies: An Exploratory Essay." *Progress in Human Geography* 24, no. 4: 591–612.

———. 2002. "Postcolonial Geographies: Survey-Explore-Review." In *Postcolonial Geographies*, edited by Alison Blunt and Cheryl McEwan. New York: Continuum.

Simon, David. 2007. "Beyond Antidevelopment: Discourses, Convergences, Practices." *Singapore Journal of Tropical Geography* 28: 205–18.

Simon, David, and Klaus Dodds. 1998. "Introduction: Rethinking Geographies of North/South Development." *Third World Quarterly* 19, no. 4: 595–606.

Sivaramakrishnan, K., and Gunnel Cederlof. 2006. "Ecological Nationalisms: Claiming Nature for Making History." In *Ecological Nationalisms*, edited by Gunnel Cederlof and K. Sivaramakrishnan. Seattle: University of Washington Press.

Slater, David. 2004. *Geopolitics and the Post-Colonial*. Oxford: Blackwell.

Solon, Pablo. 2019. "Is Vivir Bien Possible? Candid Thoughts About Systemic Alternatives." In *Climate Futures*, edited by Kum-Kum Bhavnani, John Foran, Priya A. Kurian, and Debashish Munshi, 253–62. London: Zed Books.

Spivak, Gayatri C. 1993. *Outside in the Teaching Machine*. London: Routledge.

———. 2010a. "Can the Subaltern Speak? Revised Edition." In *Can the Subaltern Speak?*, edited by R. C. Morris. New York: Columbia University Press.

———. 2010b. "In Response: Looking Back, Looking Forward." In *Can the Subaltern Speak?* edited by R. C. Morris. New York: Columbia University Press.

Srinivasan, U. Thara et al. 2008. "The Debt of Nations and the Distribution of Ecological Impacts from Human Activities." *Proceedings of the National Academy of Sciences* 105 no. 5: 1768–73.

Toal, Gerard. 1994. "Critical Geopolitics and Development Theory: Intensifying the Dialogue." *Transactions of the Institute of British Geographers* 19, no. 2: 228–33.

Vihma, Antto. 2011. "India and the Global Climate Governance: Between Principles and Pragmatism." *Journal of Environment & Development* 20, no. 1: 69–94.

Wainwright, Joel. 2010. "Climate Change, Capitalism, and the Challenge of Transdisciplinarity." *Annals of the Association of American Geographers* 100, no. 4: 983–91.

Walker, Peter A. 2003. "Reconsidering 'Regional' Political Ecologies: Toward a Political Ecology of the Rural American West." *Progress in Human Geography* 27, no. 1: 7–24.

Wani, Maknoon. 2020. "Kashmir and the Rise of Settler Colonialism: The Political and Economic Exploitation of Militarily Occupied J & K." *Himal Southasian*, September 1.

Watts, Michael, and Richard Peet. 2004. "Liberating Political Ecology." In *Liberation Ecologies*, edited by Michael Watts and Richard Peet, 2nd ed. New York: Routledge.

Williams, Glyn, and Emma Mawdsley. 2006. "Postcolonial Environmental Justice: Government and Governance in India." *Geoforum* 37: 660–70.

Zehra, Ishaal. 2019. "Nepal: A Victim of Legacy of British Colonial Times." *Modern Diplomacy*, December 20.

Ziser, Michael, and Julie Sze. 2007. "Climate Change, Environmental Aesthetics, and Global Environmental Justice Cultural Studies." *Discourse* 29, no. 2–3: 384–410.

4 Environmental justice and the right to development

The politics of scale in climate mitigation

Environmental justice (EJ) is a conceptual framework that has been invaluable in understanding the notion of climate justice in multifaceted and nuanced ways. A field of study that emerged in relation to a movement grounded in tactics of the Civil Rights Movement in the United States in the 1980s, environmental justice continues to demonstrate vibrancy and dynamism in both realms – as a social movement, an area of academic inquiry, and a paradigm for environmental discourse (Mohai, Pellow, and Roberts 2009; Taylor 2000). As an area of study, EJ started with a distributional justice focus, concerned with establishing patterns of spatial distribution of environmental bads and goods, as well as access to resources to cope with environmental risks. Environmental racism was explicitly named as a structural explanation of such distributive inequalities at that early stage (Bullard 1990). There would be much debate about whether race or class was a determining factor in shaping such/environmental distributive inequalities (Mohai, Pellow, and Roberts 2009). There have been concerns raised by some about the overwhelming dominance of a Rawlsian notion of distributive justice of goods, which was addressed by Schlosberg (2009) by bringing the notion of capabilities – drawing on Sen's (1999) ground-breaking work on reconceptualizing 'development' as 'freedom' – to EJ and the related but more bio-centric concept of ecological justice (Baxter 2005). The focus on distributive outcomes or consequential justice would be extended to concerns and studies about participatory or procedural justice and would range from concerns in inclusive consultation to meaningful engagement in decision-making processes (Mohai, Pellow, and Roberts 2009; Schlosberg 2009). EJ scholarship would be scaled up to the global context, such as in considering the triple injustice of responsibility, vulnerability, and capability in the global climate crisis. An understanding of transnational inequalities that arise from the depredations of transnational corporations on marginalized communities would emerge (Anand 2004; Roberts 2007). EJ studies in various contexts outside the United States would also emerge (e.g. Williams and Mawdsley 2006).

In terms of structural explanations for why various forms of injustice exist – whether distributive or participatory in US and/or global contexts – the political economy of neoliberal capitalism emerged as a key factor in much scholarship (Low and Gleeson 2002). Economic determinism in such structural explanations would be problematized (Kurtz 2009), and an intersectional approach characterized by the naming of racial capitalism would emerge (Pulido 2016). Meanwhile, significantly, US-based scholars recognized that EJ – although a new field of academic study drawing on a social movement by that name – is not a new phenomenon that originated in the late 20th century; rather, the creation of the nation-state of the United States was embroiled in colonial usurpation of territory and resources, and the consequent struggles for self-determination were in fact some of the oldest struggles for environmental justice (Cole and Foster 2000; Gilio-Whitaker 2019; Whyte 2018). This expansive understanding of EJ enables us to explain colonialism more broadly – in settler- as well as post- and neocolonial contexts – as a process of environmental injustice (Roos and Hunt 2010). As EJ scholarship thus expanded, the traditional focus on distributive and participatory justice concerns broadened to include notions of recognition, identity, and structural analyses beyond that of capitalism (Schlosberg 2009). With Indigeneity, race, and class being key axes for difference examined in EJ scholarship broadly, one area that remained relatively marginalized was that of gender (Buckingham and Kulcur 2009; Chiro 2008). This invisibility was attributed to the scale of the household at which gendered differences often play out but also to a broader culture of sexism and structure of patriarchy in society at large and in environmental and EJ organizations in particular (Buckingham and Kulcur 2009). Chiro (2008) made a compelling argument for a reproductive justice framework to a field of structural explanations in EJ that were overwhelmingly focused on productive justice. In a field as multi-faceted and expansive as this, there are bound to be debates for ascendancy among competing approaches. Given the plurality of this terrain, Schlosberg (2009) calls on EJ advocates and scholars to exercise bicameralism, described as the ability to recognize the value of multiple competing perspectives, as well as to exercise a politics of avowal as opposed to a politics of disavowal. Scholars have called for greater connections, linkages, and solidarity to be built among disparate struggles for justice within the United States (e.g. Sze 2020), but doing so across the North/South divide is increasingly important.

Environmental justice and the politics of scale

Environmental justice has been a compelling frame for characterizing the struggle over the atmospheric commons and is especially significant for having an orientation towards addressing the climate crisis that strives to center the marginalized and disenfranchised, unlike that of the mainstream environmental agenda. An EJ frame also effectively embodies a critique of Hardin's (1968) formulation of the commons and the assumption that the protection of the commons and justice are mutually exclusive pursuits. Justice is a subjective notion that resists universal characterization, or, for that matter, measurement (Walker 2012). It is an aspirational idea rather than a state of how things are. If the state of things is deemed unjust, people are mobilized to work towards justice.

What is deemed unjust in turn depends on one's social location, perspective, and understanding of the situation. Walker (2012) helpfully articulates a distinction between injustice and inequality – ideas about justice and injustice are inherently normative and subjective; inequality can be measured or described in various forms using various kinds of evidence. EJ is a claim-making process that utilizes particular kinds of evidence depicting particular kinds of inequalities deemed to be unjust and offering particular explanations for why things are the way they are and how they ought to change. But making claims to justice is an inherently political process (Harvey 1996; Walker 2012); therefore, claims are subject to contestation, as the previous chapters have conveyed. The EJ discourse has been employed in North/South climate debates, making claims for justice to nation-states, often on the basis of citizens with lower emissions; therefore, such claims are accountable to these citizens (Joshi 2014). In this chapter I focus on the ways in which the claims-making process of environmental justice intersects with the politics of scale, to elucidate a multi-scalar geography of climate justice claims; then I go on to deliberate on the ways in which multiple layers of injustice broadly understood impinge on the lives of the subaltern, particularly in the context of operationalizing a mainstream climate solution. Development is a theme that cuts across both processes.

Politics of scale

Similar to justice, scale is a notion that is socially constructed – and vigorously contested in the discipline of geography. On the one hand, there have been calls to "expurgate scale from the geographic vocabulary" (Marston, Jones, and Woodward 2005, 420); on the other, it has been argued that "scale is the means through which ecological (and related social and economic) change *is made* political" (Rangan and Kull 2009, 30, italics in original). Even Marston, Jones, and Woodward (2005, 426) recognized that "there is a politics to scale, and whether we engage it or abandon it can have important repercussions for social action". Scale politics is enabled by the ability of social actors to 'jump scales', that is, the ability to switch between different scales or levels in order to make political claims to overcome their marginalization and disadvantage. Neil Smith (1993, 90) described this process as a "reinscription of geographical scale. It promises not just the production of space in the abstract, but the concrete production and reproduction of geographical scale as a political strategy of resistance". Since neither scale nor spatial imaginaries are a given, but are rather actively contested and negotiated, we can see how the politics of scale can be fraught. When claims to (environmental/climate) justice are made on the basis of certain spatial/scalar configurations, we can see how these claims can be politicized as well. Walker (2009, 622) has noted that the plurality of EJ coupled with contestations over choice of scale and temporal and spatial categories leads to conflicting understandings "of the spatiality of distributional inequality" giving rise to conflicts over the spatialities of patterns of responsibility and patterns of outcome. Such scale conflicts do not occur in a vacuum but are linked to evolving power geometries, which are in turn inherently connected to struggles over access to and power over limited resources (Swyngedouw and Heynen 2004).

The centrality of the biophysical environment – often thought to be 'just' a backdrop in environmental justice struggles (Baxter 2005; Dobson 1999) – cannot be underestimated in these intersecting politics of scale, space, and justice claims. Their co-constitution is emphasized by others who see "all socio-political projects [as] ecological projects and vice versa" (Harvey 1996, 174). This dialectical relationship between socio-ecological processes and the scaling process underlying power configurations has been integral to political ecology (Neumann 2009). Specifically, posthumanist political ecology strives to recognize non-human agency in environmental and geopolitical dynamics (Sundberg 2011). This is exemplified in global climate politics where it could be argued that it is the agency of GHGs and their peculiarities that have enabled the kind of postcolonial North/South EJ politics that was outlined in the last two chapters, lending support to Neumann's (2009, 403) assertion of "the centrality and inseparability of biophysical processes in the social construction of scale". Indeed, it is the multiplicity of ways in which GHGs function and effect global climate change – and how they relate to other biophysical processes in myriad local contexts – that enables me to make the argument about how the atmospheric commons is related to a range of other commons such as forests and fossil fuels. GHGs once emitted are diffused into the global atmosphere, consequently producing a global greenhouse effect and global warming, which then affects the global water cycle and leads to various localized climate impacts all over earth. So this is a truly global-scale problem warranting an international response. But the impacts of climate change are felt by individuals and communities that are categorized and represented in imperfect ways at the international scale. Further, identifying responsibility for mitigating the problem is rendered challenging due to the multiplicity of ways in which inequities in emissions can be categorized; hence "climate politics is notably a discursive and scale-related struggle that seeks to emphasize one categorization over another" (Joshi 2015, 126).

There is a politics of scale involved in myriad responses to the predicament of devising a fair burden-sharing agreement for climate mitigation. The 'one-worldism' decried by Agarwal and Narain (1990) is one variety – analogous to the more-fashionable-in-recent-years notion of the Anthropocene (Crutzen 2002) – where its proponents unwittingly portray global climate change as a problem of a mythic monolithic humanity wreaking havoc on the global environment (Bookchin 1990). Doing so essentially homogenizes the human population in their contribution to the problem, which, because it knows no geopolitical boundaries, is considered a global problem. By inference, then, all humans – and by extension, countries – are expected to play a part in addressing the problem, apportioning responsibility based on an equality principle, although in reality, not all humans and/or countries have contributed in uniform proportion to the problem. A rescaling of such an issue from the global level to the level of the individual – that is, 'scale jumping' from global to individual – prompts an individualization of responsibility (Maniates 2001) and simultaneously feeds into neo-Malthusian fears of countries such as India and China with large populations as the most potential threats to climate change. In turn, individuals in these countries have engaged in their own version of a politics of scale, which is to problematize 'one-worldism' as a Northern environmental agenda, and to juxtapose this against a particular set of claims to international environmental justice represented by a common Global South position. In this equity-oriented politics of scale, the Global North and South are seen as legitimate categories along which to conceptualize and apportion responsibility to mitigate, based on evidence of disproportionate contributions, as well as that of impact, vulnerability, and capability. In addition to the strategic essentialism of the Global South – a spatial imaginary that conflates a heterogeneous community that includes emerging economies and most vulnerable countries into one negotiating bloc – this is also a state-centric imaginary that enables state representatives to 'jump scale' (Smith 1993) when emphasizing a per capita rights approach to environmental space or the global atmospheric commons, and then to extrapolate this right from an individual level to the state, rationalizing why the Indian state should not be subjected to caps on GHG emissions. This is particularly problematic since the Indian state has been found guilty of multiple environmental injustices with respect to distributive, participatory, and recognition dimensions (Williams and Mawdsley 2006), which could be read as a contradiction, given its discourse on climate equity on the international front. Even as I have argued and continue to emphasize that the strategic essentialism and 'scale jumping' of Indian elites should not be a priori problematized, here I want to note that such scale jumping raises expectations for accountability in domestic and regional contexts, and in the case of India, such discourses occur at the intersection of climate justice claims and discourse of development (Joshi 2014).

In reflecting on the ways in which the politics of scale and politics of spatial imaginaries intersect in global climate politics, two observations are warranted. Firstly, the nature of the problem makes certain kinds of politics possible. This includes, as earlier noted, the key role of non-human nature in engendering certain political possibilities to which we should remain open. At the global scale, this might mean a resurgence of North/South justice claims. At local scales, it might mean more control of communities over their resource commons due to the unpredictability of ecological change and the need for adaptive governance (Ostrom 2010). Secondly, environmental or climate justice claims, as Walker (2012) has argued, are based on specific kinds of evidence about the nature of inequality and how they are categorized and presented. Here, it is not only spatial scale but temporal scale also that is relevant, and both are subject to contestation. Contributions to GHG levels in the atmosphere can be categorized according to several criteria, and one aspect of climate politics is to emphasize one over another. A side event during COP-15 hosted by the Chinese government in the Bella Center emphasized a cumulative per capita equity argument, according to which China's and India's contributions are significantly less than that of the United States, relative to comparing current per capita emissions, whereas if the focus is shifted to total country contributions, China, the United States, and India lead the world in that order. India's INDC (n.d., 2) notes:

> The cumulative accumulation of greenhouse gases (GHGs) historically since industrial revolution has resulted in the current problem of global warming. This is further compounded by the tepid and inadequate response of the developed countries even after the adoption of the United Nations Framework Convention on Climate Change (UNFCCC) and delineation of obligations and responsibilities. As a result, an "emission" ambition gap has been created calling for enhanced global actions to address it. India, even though not a part of the problem, has been an active and constructive participant in the search for solutions. Even now, when the per capita emissions of many developed countries vary between 7 to15 metric tonnes, the per capita emissions in India were only about 1.56 metric tonnes in 2010.

The emerging emitter discourse, on the other hand, emphasizes these two countries' total emissions and projected future emissions.

My interviews with Indian officials reflected a similar emphasis on cumulative emissions:

> At an international level, environmental justice is Annex I countries stepping up to the plate and taking the lead with dealing with the problem that they created . . . Since they are responsible for creating carbon stock, they are responsible for taking the lead.
>
> (Interview, male, Bureau of Energy Efficiency, Delhi, December 2008)

> When you look at climate change and what is causing it, we should look at CO2 stocks in the atmosphere, and not just flows. . . . We should look at the portion of atmospheric space occupied by historical and current emissions of a country.
>
> (Interview, male, MoEF official, Delhi, December 2008)

Climate science weighs in favor of the cumulative responsibility argument, since the long life span of the most common GHG, carbon dioxide, renders the role of past emissions crucial (IPCC 2014). But the spatial categories apportioning cumulative responsibility along North/South lines have been vigorously contested as already described. Further, the 'territorial trap' argument questioning states as the appropriate container for the comparison of emissions profiles has been raised to argue for shifting responsibility from countries to corporations, specifically in the fossil fuel industry (Griffin 2017). In my teaching as well, I have encountered much discomfort (among US students) with the per capita approach and have found a tendency among students to want to deflect attention away from data scaled to 'the average citizen' and to focus on corporate or military emissions. Even if such deflection of responsibility seems problematic, it is indeed important to recognize that such emissions data averaged across a country's population can significantly distort ideas of responsibility, especially in countries where disparities in income/emissions are extreme. Ananthapadmanabhan, Srinivas, and Gopal (2007) had on similar grounds problematized the scale-jumping in India's stance on climate by claiming it was "hiding behind the poor", in demanding to be absolved of mitigation responsibilities, by basing its arguments in international negotiations on its overwhelmingly poor citizens whose lack of access to energy and other basic needs helped bring the average per capita emissions down. Their study disaggregated per capita emissions in India for three income groups. Even so, the per capita emissions of India's richest 10 percent were found to be less than a tenth of the per capita emissions of the richest 10 percent in the United States and were the same or less than that of even the poorest 10 percent in the United States (Narain 2019). As such, many in India share a keen sense that Annex I countries have instead hidden behind the emerging economies as a delaying tactic and to conceal their own failures in meeting their mitigation responsibilities. Rather than invalidating the official Indian position on climate negotiations, the key idea in the 2007 Greenpeace India report was to *also* hold the

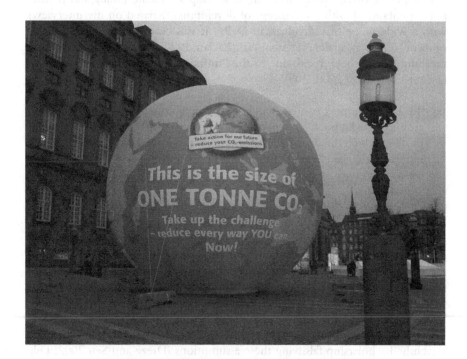

Figure 4.1 Art installation in Copenhagen during COP-15, December 2009

government accountable to its politics of scale. Taking such complexities into account, Baer et al. (2009) developed a sophisticated plan for a fair burden-sharing of mitigation responsibility that took into account emissions at the scale of nation-state, income groups, as well as sectoral emissions, but such proposals have not yet been taken up in climate negotiations.

Development as 'Floating Signifier' in climate discourse

One of the key positions of the Global South in climate discourse – as reflected in the UNFCCC, Kyoto Protocol, and the Paris Agreement – is the 'right to development' of the Global South. This right is claimed in light of both the North's disproportionate usurpation of limited ecological space historically from the colonial era to the present and the perceived constraints to pursuit of development due to shrinking ecological (i.e. atmospheric) space. Clearly development aspirations and ecological concerns are seen as inherently connected. But how are the two connected in the specific context of climate change and climate justice? When a 'right to development' is asserted, what is meant by development? Is there a politics of development, and if so, (how) does it intersect with the politics of scale, space, and justice? What are the implications of a 'right to development' approach to climate justice for the issue of domestic inequities within states and by extension within the Global South? These are important questions to heed so long as climate policy and politics remain entangled with the concept of development. Based on my interviews with a wide range of individuals in Delhi, it was clear that the term 'development' serves as a 'floating signifier' (Laclau 1990, 28) – a concept whose meaning has not been fixed, due to the "unfixity introduced by a plurality of discourses" (Laclau 2000, 305).

> To 'hegemonize' a content would therefore amount to fixing its meaning around a nodal point. . . . periods of 'organic crisis' [are] characterized as those in which the basic hegemonic articulations weaken and an increasing number of social elements assume the character of floating signifiers.
>
> (Laclau 1990, 28)

Climate change is clearly an organic crisis that has created an opening for a recalibration of what development should mean. As established, the contemporary North/South politics has not successfully created a counterhegemonic decolonial politics yet, and the possibilities lie in whether and to what extent the notion of development could create a discursive space for the creation of such a counterhegemony to the coloniality of power.

The architecture of the global climate agreements associated with the UNFCCC has consistently presumed the importance of economic growth and sustainable development with well-being in developing countries, despite an established scholarship falsifying these assumptions (Dreze and Sen 2002; Lele 1991). This was also evident in India's climate policy discourse: "Maintaining a high growth rate is essential for increasing living standards of the vast majority of our people and reducing their vulnerability to the impacts of climate change" (GoI 2008, 2). The report goes on to argue that "large-scale investment of resources in infrastructure, technology and access to energy" will be required to meet this goal. Further, "only rapid and sustainable development can generate the required financial, technological and human resources" required for climate adaptation (ibid, 13). These sentiments have remained in India's INDC submission to the Paris Agreement and are reminiscent of NIEO proposals that sought to repair the unfair terms of trade between North and South by calling for self-sustaining economic growth and industrialization (Bhagwati 1977). Within those participating in climate discourse in Delhi, the notion of 'development' carried different meanings – economic growth, well-being, and capability to cope with climate change – with clear connections imagined between them. When 'right to development' was equated with 'right to economic growth', this right was typically associated with the nation-state; on the other hand, when 'right to development' was described as well-being, it was typically associated with the average citizen in India (Joshi 2014). It is this flexibility of meaning that enables the 'scale jumping' between individual and state described earlier.

The climate crisis was seen as an opportunity to balance the status quo and to bring countries – and by extension, their citizens – at par with one another. If India as a 'developing country' prematurely accepted mitigation responsibility not in line with its historical responsibilities and capabilities, there was a sense that the setback would be felt by its citizens: "We will not agree to a concept of a peaking year for India, because we have a huge backlog of development still awaiting us, particularly in expanding rural electricity supply" (Participant observation, Jai Ram Ramesh, then Minister of Environment and Forests, December 2009, Copenhagen). The perceptions that poorer countries are going to be the worst sufferers of climate change and that India's socioeconomic vulnerability was caused by the Global North were particularly strong. A strong link was generally perceived between GHG mitigation commitments and India's ability to alleviate poverty, which were seen as mutually exclusive; but international climate justice claims made by India were seen to be compatible with the task of attending to intra-national disparities in development. If India were to take a bigger share of emission targets of India than is warranted, it was presumed that this unfair burden would be shouldered by India's poor rather than the rich. The point made frequently was, why should an average Indian citizen that had no role to play in the creation of the problem have an obligation to contribute to the solution by compromising his or her right to economic development?

See the basic premise of the climate debate has been on right to development and right to equality and growth and a better lifestyle. Sitting in US, and in other countries, without doing anything on your own, it's much easy to say that okay, India and China should be seen differently, right? It's easy to say that. Why should you see India and China differently, is my question. What India and China is trying to do is improve the quality of life of the people! I mean, . . . if you look into numbers much more clearly you will find large number of much more poor people in India, still, without access to even basic facilities. . . . should these [countries] therefore stop providing access to better quality of life to these people?

(Interview, MoEF official, male, December 2008, Delhi)

Developing means, a level of development required to serve basic needs. Through industrialization, education, domestic energy consumption . . . all these need consumption of energy and GHGs. When it comes to climate change it is the carbon space we are talking about. . . . The word developing does not necessarily imply that a certain goal of being like a certain set of countries is being pursued. Every country can decide what its development goals are. For India, it is about attaining a basic standard of living, access to health care, education, drinking water, etc. it's not that it is trying to be like the West. Not at all!

(Interview, MoEF official, female, December 2008, Delhi)

Economic development was seen as not only necessary for poverty alleviation but to enhance the state's adaptive capacities including infrastructure for adaptation to climate change. Vulnerability was understood not simply as an outcome of the impacts of climate change, but this combined with the socio–economic capabilities to cope with these impacts. A senior TERI official who has served as a long-time climate negotiator for India argued that the focus for developing countries like India – that is also highly vulnerable to climate change and has a sizable population that is economically poor and/or dependent on monsoon-fed agriculture – should be on enhancing adaptive capacities rather than on mitigation. The only way to enhance these capacities was seen to be through rapid and sustained economic development (Interview, January 2009, Delhi). As with the GoI report earlier, a strong economy was perceived to be a prerequisite to greater economic well-being for ordinary citizens; and relatedly, state capabilities to withstand climate shock were seen as integral for citizens' capabilities to do so. The government was perceived as having a responsibility to ensure the well-being of its citizens. In other words, the dual roles of the state – pursuing economic growth and industrialization and attending to the needs of the citizens – were seen as complementary, and the state was in fact viewed as an arbiter in facilitating an equalization in living standards in the world that were perceived to be blatantly unequal. Several interviewees remarked on the Indian state's accountability towards its citizens as a redistributive justice imperative, thus establishing a direct link between national mitigation commitments and the individual Indian citizen, and validating the state-centrism in North/South differences. In this sense, the 'scale jumping' politics of the Indian state in climate negotiations enjoy a degree of legitimacy among the Indian intelligentsia, even as there are frequent contestations and debates between the state and civil society (Joshi 2014). Commenting on the question of EJ within India, a civil society member said:

> I think equity within nations is something that is being ignored. . . . For instance, a recent report says that 70% of Indian population survives on 20 rupees a day income. Now despite the emerging economy story and high growth rates, if that is the case then one really needs to look at what is happening and how it is happening. And because they are poor, their vulnerability to climate change is also high. The interesting thing is that the climate change and poverty linkages are very complicated. . . . It's right that we are growing very fast or we are, in some sectors, in some areas competitive, but the fact remains that a large part of our population still lacks basic facilities. So we need to provide them electricity, we need to provide them roads, we need to provide them good houses, we need to provide them better education, and transport and other services. So we can't really take the emission reduction targets. . . . So that's the complicated link, but then the question is whether despite all that whether . . . the development that has happened in last decade has actually gone to those people, that's the question . . . I feel that the key to the ethical approach to any subject is consistency, at every rate.
>
> (Interview, TERI official, male, December 2008, Delhi)

In a follow-up interview a year later, he was similarly critical of his government's narrow conceptualization of development to the detriment of its most vulnerable citizens: "the biggest opportunity that climate change has provided is to reform, revitalize the overall development agenda . . . [but] the same issues that are being raised at international arena are not getting same attention or same emphasis at domestic level" (ibid, December 2009, Copenhagen). This was an argument for holding government representatives who promote a North/South equity agenda in the international sphere accountable to their 'scale jumping'. Similar arguments were made by individuals from Bangladesh and Nepal whom I interviewed at the COP-15 in Copenhagen. While my research in India was not focused on any specific development project, published accounts give a sense that there is indeed a disconnect between development discourse and practice (Agarwal et al. 1999; Ananthapadmanabhan, Srinivas, and Gopal 2007; Lele 1991; Williams and Mawdsley 2006). Nevertheless, the idea of development as a key priority is prevalent beyond the class of government representatives. An Indian Youth Climate Network (IYCN) member giving his point of view on how development was central to climate debates and why questioning India's right to development is unwarranted, shared:

> See I think if we look at everything from a development point of view, everything, we would be [having] a completely different debate altogether, we'd be discussing how to bring the countries who are less developed right

now at the same level. So personally, I would say, let's answer questions on a developmental index. Not on climate change index, not on emissions index, let's debate on the development index itself. . . . That's my personal standpoint on the issue, and you know, because of climate change what has happened is that development itself has become a hindrance, in some way. I mean, if somebody asks what's the biggest challenge India faces right now, I would say it's development. Because it's such an important thing you need to develop in the right direction . . . no doubt, but you have to define development. And that probably is the biggest challenge. How do you define development? Because you know it has different meaning for different set of people.

> (Interview, IYCN member, male, December 2008, Delhi)

Individuals I spoke to were keenly aware of (mis)perceptions outside India that a right to development was understood in India as a right to pollute; they were eager to set the record straight – India was keen on leap-frogging past coal and other dirty sources of energy development to pursue an ecologically benign economic growth development paradigm that was delinked from emissions, with the help of financial and technological transfers from the Global North. As such, they were eager to distinguish India's aspirations for development from the model pursued by industrialized countries. Yet, industrialization and economic growth were often assumed to be a necessary condition for alleviating poverty and meeting energy and basic needs for the poorest and rarely were they seen to be competing interests. This framing of the desire for industrial growth in the language of basic needs is interesting when seen in the context of the US agenda of advancing a similar discourse alongside the active rejection of calls for an NIEO in the 1970s (Hansen 1980). Whereas the US response to the NIEO was development-as-aid with a focus on meeting basic needs, the reincarnated struggle for economic justice in the context of climate debates enables India to assert self-determination in attending to its citizens' basic needs by being a stronger economy. No matter whether the meaning of development was associated with economic growth, well-being, or capability enhancement, a striking observation is that all of these pursuits were typically constrained by a neoliberal and modernist paradigm – if sustainability or sustainable development is the goal, then it could be pursued via ecological modernization (Joshi 2014). On their part, Indian officials see the neoliberal constraints as a vestige of neocolonialism:

> For decades, we have been lectured to by the multilateral financial institutions about growth being the necessary precondition for poverty alleviation. . . . If we have to put curbs on GHG emissions given that at present we don't have clean technologies that are anywhere close to the costs and the scale of fossil fuels, it means that we are clamping down on our energy

use and thereby our economic growth. And in that circumstance, it is not possible for us to alleviate poverty in our countries.

(Interview, Distinguished Fellow TERI, male, December 2008, Delhi)

This overwhelming dependence and faith on neoliberalism seems unfortunate, particularly in light of the contributions of key thinkers from the region, such as Sen (1999) in reconceptualizing development. Challenging the dominance of Rawlsian (distributive) justice norms in mainstream development theory – particularly those that equate development with income – he had argued for a novel conceptualization of development as the enhancement of freedoms. This work has been quite influential in helping EJ scholars recalibrate a similar focus on distributive justice elements in EJ theorizing. Drawing largely on Sen's work, Schlosberg (2009) argued for EJ to be understood in terms of capabilities – that is, to study EJ struggles in terms of whether the socio-ecological capabilities of an individual or community are being thwarted or supported by prevailing circumstances. While I do not support anti-modernist de-growth arguments unilaterally imposed on the Global South from a positionality of Global North privilege, I do believe the idea of viewing 'right to development' through a lens of capabilities is a promising area of investigation for scholars of development and environmental/climate justice and that such a focus on capabilities would help navigate the relationship between development rights conceived at different scales. Such a reformulation of the meaning of development is critical for counterhegemonic North/South struggles, by decolonizing mainstream development which tends to be techno-centric and overwhelmingly focused on economic growth. The organic crisis of climate change provides a strong rationale for 'right to development' claims to be scaled to communities:

. . . the concerns of the poor and marginalised. These concerns are poorly represented by Southern governments and civil society, and are often lost in the plethora of Northern perspectives that dominate the global agenda. Global negotiations may seem currently far removed from local and more immediate Southern concerns, but ignoring them in these early stages, when the rules are being set for future global governance, would be tantamount to the people of the developing world giving up their rights as global citizens. Southern actors not only have to be active and alert participants in the process, they also have to be aware of their responsibility to ensure that global decisions reflect local concerns, and of their responsibility to future generations, who will have to live with the rules that are made today. The current trends of negotiations reflect mostly Northern concerns, and have slowly turned into business transactions rather than systems of governance based on democracy, equality and justice.

(Agarwal et al. 1999, v–vi)

Clean development mechanism in the Kyoto Protocol

The Kyoto Protocol not only introduced the Annex I/non-Annex I distinction for determining mitigation responsibility. In the process of accommodating US priorities, the treaty included a cap-and-trade provision for GHGs, based on a similar provision successfully instituted for sulphur dioxide emissions in the United States (Ervine 2018). Introduced in 1997, cap-and-trade was a market-based mechanism meant to help countries fulfil their mitigation commitments under the Kyoto Protocol by sponsoring carbon mitigation elsewhere. The Clean Development Mechanism (CDM) was one of three such flexibility mechanisms that enabled trading of emission offsets across the Annex I-non-Annex I divide. Annex I countries that were assigned a mandatory ceiling on GHG emissions under the Kyoto Protocol could in theory adhere to the designated limit by purchasing certified emission reductions (CERs) from non-Annex I countries that did not have mandatory reduction targets and were interested in undertaking voluntary emission reduction projects. Under the CDM, developing countries can earn CER credits, each equivalent to one ton of carbon dioxide, following an elaborate vetting and certification elaborate process by the UNFCCC's CDM Executive Board. Two essential criteria project proponents must meet are verifiable, certified emissions reductions documenting mitigation potential and locally relevant sustainable development. Tied to the first criteria is additionality, which is the need to document verifiable emission reductions tied to the project, which would not materialize in the absence of CDM. The price of CERs ranges from USD 0.4 to 5 per CER. A total of 7,833 CDM projects have been registered by the UNFCCC so far, accounting for more than two billion CERs. A portion – 2 percent – of the income generated from trading CERs contributes to the Adaptation Fund to finance adaptation efforts in the most vulnerable developing countries (UNFCCC 2020). With the Kyoto Protocol set to expire, the CDM's future is uncertain, although CERs issued under the Kyoto Protocol should extend to 2022 under the Doha Amendment. Subsequently it may be replaced by the provisions of Article Six of the Paris Agreement in the form of the Sustainable Development Mechanism (ibid).

Carbon trading, or carbon offsetting, has been heavily criticized for a number of reasons, including its inefficacy in reducing GHGs globally due to loopholes buyers and sellers are able to exploit; the tendency to reward polluters due to the additionality requirement; the fluctuations in price of carbon; the problem of incommensurability caused by the different characteristics of GHGs and the varied contexts in which their curtailment is being attempted and where they will be counted, across the luxury-subsistence emissions divide; as well as the opaqueness and complexities of such quantifications and trade, making meaningful participation by affected communities practically impossible (Ervine 2018). It has also facilitated the commodification of the atmosphere (Barnhart 2013), thus adding yet another layer to the paradigm of nature's neoliberalization (Bakker 2005). As with other forms of international trade, the carbon market, too, is situated within the context of core-periphery power dynamics. As such, CDM has been critiqued due to claims of 'carbon colonialism' and 'green capitalism'. Bumpus and Liverman (2010, 204) have argued for a multi-scalar analysis of the relative and relational political economy dimensions of these carbon offsets "as a new commodity that links north and south through a complex set of technologies, institutions and discourses". Key aspects to be examined in assessing the promised win-win outcomes of climate mitigation and sustainable development from carbon trading include "the extent to which a carbon market can realistically attend to certain forms of development, the limitations of market mechanisms in governing emissions reductions more generally, and the crucial questions of power relations in North/South (and South/South) environments and development outcomes as a result of offsetting and wider flows of capital for clean or dirty development" (ibid, 219).

Now, as with 'development', 'sustainable development' has been an incredibly slippery concept to pin down and may be referred to as an empty or floating signifier (Laclau 1990, 2000). The most commonly used definition of sustainable development has been offered by the Brundtland report: "Humanity has the ability to make development sustainable to ensure that it meets the needs of the present without compromising the ability of future generations to meet their own needs" (Brundtland 1987, 7). A key critique of this definition is its vagueness, particularly in the absence of a clear understanding as to what is to be sustained, how, for whom, and to what end. Another critique is the close association of the concept with economic growth, even as the need for a global conversation about sustainability and sustainable development had risen in response to idea of the 'Limits to Growth' (Lele 1991; Meadows et al 1972). The Brundtland report, while acknowledging that growth could not continue forever in a finite world, emphasized the continued necessity of economic growth to address Third World poverty and basic needs but committed to supporting 'a different kind of growth' facilitated by ecological modernization. The economic determinism and focus on techno-fixes at the cost of participatory and locally meaningful development would be problematized by many scholars, and sustainable development as it had been predominantly practiced would be relegated to the status of an oxymoron (Redclift 2005). Yet the term features prominently in the Paris Agreement in the context of both mitigation and adaption efforts. As Lele (1991, 19) remarked,

> if SD is to be really "sustained" as a development paradigm, two apparently divergent efforts are called for: making SD more precise in its conceptual underpinnings, while allowing more flexibility and diversity of approaches in developing strategies that might lead to a society living in harmony with the environment and with itself.

Figure 4.2 A household biogas plant in Gorkha, August 2017

Unpacking neoliberal climate solutions and 'women in development' – Nepal case study

The CDM has long been critiqued for being ensconced within this neoliberal paradigm of sustainable development, but also for its propensity to reward polluters and large-scale development projects. Of the 12,000 registered projects, most have been located in China, India, Mexico, and Brazil. Within Asia, India and China – the emerging economies that have most ardently presented themselves as developing countries in climate negotiations – have received the biggest share of CDM-generated funds, with approximately 70 percent and 15 percent of registered CERs, respectively (UNEP/DTU 2020), going to show that the economic benefits from the cap-and-trade provision of the Kyoto Protocol have been unequally distributed within the Global South, in addition to perpetuating unequal terms of trade between the Global North and South. The approved projects have predominantly been large-scale hydo-power projects, clean coal projects, or industrial scale waste to energy conversion projects (Bumpus and Liverman 2010; CDM Watch 2012). Apparently, then, sustainable development has been envisioned primarily from an ecological modernization perspective that sees pursuit of sustainability primarily as a technological challenge. In several cases CDM projects have led to the disenfranchisement of Indigenous populations who relied on project-affected land (e.g. being submerged by dams) for their subsistence and cultural needs (CDM Watch 2012; Finley-Brook and Thomas 2011). Bumpus and Liverman (2010) have suggested that CDM funds should be channeled to smaller-scale projects that may have greater potential to support more meaningful sustainable development. Here I share my observations based on fieldwork in Kathmandu, Lalitpur, and Kabhrepalanchowk districts of Nepal during August 2013 to examine a smaller-scale CDM project to better understand how claims to meaningful sustainable development are made in the local context, drawing on interviews with Kathmandu-based officials and project beneficiaries at the household level, as well as analysis of textual documents. Nepal had pursued and was approved for six CDM projects in the category of small-scale projects, of which four were for methane avoidance from the adoption of biogas technology at the household level (UNFCCC 2020).

Renewable energy as 'clean development' in Nepal: Biogas

Commonly referred to as an appropriate technology for renewable energy production, biogas – or *gobar gyas* in the local vernacular – is a methane-rich and flammable gas produced from the anaerobic digestion of cow manure and other organic biomass and can be used as a cooking fuel. The technology was introduced by the Netherlands Development Organization in Nepal (SNV/N) in the 1950s and was actively promoted since the 1970s by the Government of Nepal (GoN) and non-governmental agencies to enhance the rural population's self-reliance for energy and to reduce dependency on India for petroleum imports. The Alternative Energy Promotion Center (AEPC), the designated national authority (DNA) for CDM implementation in Nepal – affiliated to the Ministry of Science, Technology, and Environment (MoSTE), now Ministry of Energy, Water Resources and Irrigation (MoEWRI) – successfully secured CDM project funding for biogas projects for the first time in 2003. Today the AEPC facilitates eight CDM projects and programs including household Biogas and Micro-hydro projects, which have generated 2.9 million CERs that have been sold for USD 14.42 million. By 2019, AEPC had facilitated the installation of 425,511 domestic biogas units (GoN 2019), of which 243,882 were supported by CDM since its start in 2003 (Email correspondence with AEPC, August 2020). The existence of long-term baseline data in rural household energy usage has aided in the documentation of the much-needed additionality requirement. Emission reductions are attributed to reduced consumption of firewood – which constitutes non-renewable biomass and is the primary source of energy for cooking in rural Nepal – when it is substituted by biogas. An indirect benefit is conservation of forests, a vital sink for carbon dioxide. Barnhart (2013) noted that with the advent of the carbon market in Nepal, the governance of biogas has entered a neoliberal phase, especially since donor funding for development was retracted in the 2000s.

Since its inception as a development project, biogas technology was subsidized by the Nepalese government – first with donor agency support, and now with carbon financing – such that an individual household would pay a reduced cost of approximately NPR 42,200 (NPR 34,000 if contributing labor for installation) for a biogas digester costing NPR 65,000 (as per an ER Rights Transfer for Biogas contract, dated September 1, 2018). In exchange for the subsidy received, amounting to NPR 22,800, the recipient of the subsidy relinquishes their carbon rights when signing a contract, without being informed about this based on the presumption that they wouldn't understand (Barnhart 2013). A 2017 Project Design Document prepared for the AEPC states in no uncertain terms that when a household receives a subsidy for a digester, they are essentially transferring ownership to 'emission reduction' carbon credits generated from it to the AEPC:

> The owners of a digester signed an agreement with AEPC by transferring all legal rights, interests, credits, entitlements, benefits or allowances arising from or in connection with any greenhouse gas emissions reductions arising from the operation of the digester (Emission Reduction), and agrees to take all necessary action required to ensure the transfer of those Emission Reductions to the Alternative Energy Promotion Centre or its nominee, including executing any relevant documents.
>
> (PDD 2017, 6)

An Emission Reduction (ER) Right Transfer for Biogas contract, dated September 1, 2018, titled (translated from Nepali) *Government of Nepal, Ministry of Population and Environment, AEPC, Household Biogas Plant Subsidy Application Form*, with the beneficiary's fingerprints, included a Nepali translation of the previous statement about giving up legal rights and benefits emanating from GHG emission reductions from the biogas plant.

Here the AEPC essentially serves as an intermediary between the households that enable emission reductions and companies interested in buying these CER credits, brokered by the World Bank, with whom the AEPC has signed an ER Purchase Agreement. The CERs produced by individual households are pooled together and sold by the AEPC to the World Bank (Barnhart 2013). Although technically the seller of CERs, the individual household does not benefit directly from the sale – except for the subsidy amount – although in theory the benefits are passed on to other households, since the AEPC claims to reinvest the CDM-generated funds to subsidize future biogas plant installations elsewhere. During 2018–9, AEPC generated USD 2.29 million on the basis of creating 0.6 million CERs (GoN 2019). This is seen as crucial to sustain the biogas program at a time when foreign aid for development is dwindling. Given the question asked earlier of 'what is to be sustained', it is clear here that the institutional infrastructure of biogas technology is the object to be sustained. But is this how the 'sustainable development' criteria is met for CDM certification?

Women in (sustainable) development

Ministry and AEPC officials revealed during interviews that the criteria for sustainable development for biogas-as-CDM project are met by improving the lives of women, particularly by enhancing their access to affordable, clean, and modern renewable energy. Women's lives are thought to improve in several ways, all connected to the reduced reliance on firewood that is achieved by switching to biogas: they no longer have to walk long distances to collect firewood from the forest or pay ever-rising prices for them due to increased scarcity – the saved time and resources could be invested in economically productive and socially meaningful activities; and they have less exposure to indoor air pollution which is a significant source of mortality for women and children in rural Nepal – over 7,500 women and children in Nepal are estimated to die every year due to smoke-induced illness caused by use of traditional cook stoves (NESS 2011; PDD 2017). The annual review document notes that biogas supports Sustainable Development Goals pertaining to good health and well-being, affordable and clean energy, and climate action (GoN 2019). The women I interviewed reported positive experiences with this technology. It did not add to their burden of work. The biogas cook-stoves were just as easy to operate as kerosene gas stoves, and easier to operate than the traditional cook-stove using firewood. In all households I visited in Godavari (Lalitpur district) and Sanga (Kabhre district), all three ways of cooking were followed – the adoption of biogas had not replaced other means of cooking. Most of the women found the subsidized cost to be reasonable – one of the four households I visited in Sanga could not afford even the subsidized cost given other competing expenses. In two of the households, the slurry from the biogas plant was used to fertilize the *karesa bari*, or kitchen garden. One household had one functioning unit and was in the process of installing a second one to be attached to an outdoor toilet.

An interesting anecdote from one of my visits illustrated a gendered division of labor that I was myself all too familiar with and that critical geographers have noted in the patriarchal context of South Asia (e.g. Sultana 2013). While visiting a household, I spoke with the male head of the household, while his wife was away. After a brief conversation, he enthusiastically proceeded to show me how the biogas stove operated. Neither he nor another male (tenant) in the premises was able to turn the stove on even after several attempts. They were both flustered, having assumed it was easy to do! In a little while, the wife returned and promptly turned the stove on. It was clear from conversations with female members of this and other households I visited that all the work associated with operating the biogas stove belonged to the woman – collecting and transporting manure; stirring the sludge periodically; extracting fertilizer from the slurry and applying it in the *karesa bari*; and, not surprisingly, using the stove to cook with. In each of the households, although it was clear that it was the woman's labor that makes possible this particular carbon trade between the AEPC and the WB, it was the male head-of-household who made the

Figure 4.3 An outdoor toilet under construction, to be connected to the biogas stove in the kitchen (overleaf), Sanga, August 2013

decision to adopt the technology. As indicated earlier, this decision entails signing a contract stating that the household voluntarily agreed to receive a subsidy from the AEPC, and in return agreed to relinquish any future claims from the transaction. As Barnhart (2013) noted, these decisions have not been made by rural households based on an understanding of how they fit into the global trading apparatus.

The absence of fully informed consent is a reflection of the loss of agency of the household, and particularly, of the woman, whose labor makes this transaction possible. The way in which the 'sustainable development' criteria for this program is met reveals a 'women in development' approach decried by feminist development scholars because it regards women simply as victims needing to be rescued and not as beings whose agency matters (Sheppard et al. 2009). In this case, 'sustainable development' entails the empowerment of women by rescuing them from indoor air pollution and the drudgery of firewood collection, without seeking to problematize gendered power dynamics in the household, or seeking to enhance women's agency in decision-making that affects their lives – for instance, by informing the design of biogas plants, or suggesting a graduated subsidy – or seeking to ensure that they are compensated fairly for their reproductive labor (Joshi Forthcoming). The PDD (2017, 10)

Figure 4.4 Trying to light the biogas stove, Sanga, August 2013

tellingly argues that the biogas CDM program meets 'gender sensitiveness' criteria because of the acknowledgment that "gender relations, roles and responsibilities exercise important influence on women and men's access to and control over natural resources and the goods and services they provide, the project has given access to biogas to both men and women without inequality". The annual review document similarly notes the internalization of 'gender and social inclusion' activities throughout AEPC programming (GoN 2019). Yet interestingly, even as women are singled out for being the beneficiaries of this technology, they were largely absent in positions of authority in the CDM/AEPC institutional architecture, as also evidenced by my predominantly male interviewees.

Mitigating global climate change with cow dung?

The climate mitigation potential of biogas projects relies on the claim that the diffusion of this technology effectively replaces the use of non-renewable energy sources – predominantly firewood but to a lesser extent fossil fuels such as kerosene and liquefied petroleum gas (LPG) – with renewable energy (Dhakal and Raut 2010; NESS 2011). The mitigation benefit is based on the conversion of methane to carbon dioxide when burning – carbon dioxide is less potent than methane as a GHG – and the fact that less carbon is generated than from burning firewood (Barnhart 2013). While project proponents argue that the additionality requirements of CDM are met by ensuring that emission reductions are "claimed only for the displacement of non renewable fuelwood" (PDD 2017, 3), it is hard to imagine that the displacement is complete. In the households I visited in 2013, women shared with me that the flame of the biogas stove was not big enough to cook meals and that they mostly used it to make tea, and to boil milk and water. Firewood and LPG gas cylinders would frequently supplement the use of biogas. But passing judgment on the everyday household practices of economically impoverished rural households to resolve a global climate crisis caused by industries and consumers in the Global North hardly seems conducive to a climate justice imperative, no matter how you define it. Doing so implies that while luxury emissions can go unchecked, subsistence emissions will be policed. My students are often enraged at this perceived injustice, even as they (or any of us) may be equally complicit and powerless when we participate in activities such as flights, where claims are made about carbon neutrality.

Of all the 'unclean' development practices that currently litter the Nepalese landscape, why did biogas emerge as top contender for CDM? The challenge posed by the additionality constraint was named by several officials I spoke to. Even government insiders find the idea of mitigating climate change with biogas – in a country with insignificant per capita emissions – incongruous, even as they welcome the financial flows it makes possible. They argue that had it not been for the additionality requirement, much more effective mitigation strategies, even within Nepal, could be supported. Relatively larger-scale emission reductions could be achieved, for instance, by reforming the transportation sector and targeting the highly polluting brick industry (see also Dhakal and Raut 2010). But the process of seeking CDM projects has been challenging to navigate for Nepalese officials who lamented the lack of capacity and expertise in navigating the complexities of global carbon trading. Consequently, a heavy reliance on foreign consultants for preparation of the proposal design and development (PDD), certification, as well as MRV, has led to the siphoning away of much of the financial resources gained through the CDM, rendering most CDM projects economically unfeasible (Interview, AEPC official, male, August 2013, Kathmandu). Officials were also wary of investing time and resources on CDM feasibility studies when the Kyoto Protocol itself was slated to end in 2020. But new CDM projects and programs are in the pipeline and the AEPC, having become accredited by the Green Climate Fund (GCF) as the direct access entity (DAE) for funding, is seeking to register additional projects, such as industrial scale biogas, micro-hydropower projects, as it strives to scale up renewable energy to achieve 12 percent of the country's energy use by 2024 (GoN 2019).

It is clear that the implementation of biogas as a climate solution is riven by multiple inequities at various sites and scales. My interviews revealed that the flat subsidy was unhelpful to the most economically constrained households. Access appears to be governed by caste, with privileged caste groups being the primary beneficiaries: a 2010 survey showed that Bahun/Chhetris constituted

60.75 percent of biogas adopters, Adibasi/Janajati 10.37 percent, and Dalits represented only 2.96 percent (NESS 2011). As with the approach to gender-sensitization, efforts to address this disparity are limited to 'targeting' Dalit and Adibasi/Janajati groups to become beneficiaries. Besides the gendered nature of burdens and benefits discussed earlier, gender and caste disparities were also visible in salaried positions of the institutional architecture within which CDM and biogas projects operate (GoN 2019). Then there is the urban/rural divide between those who are project 'targets' in firewood-dependent rural house-holds and those who are decision-makers and implementers employed in gov-ernment and non-governmental organizations, with formal English-medium education, typically based in Kathmandu and other cities. Thus within Nepal, inequities exist along lines of geography, bureaucratic/technocratic authority, caste/Indigeneity, income, and gender. Scaling up from the nation-state, we know that benefits are inequitably distributed within the Global South. The unequal North/South status quo is maintained by the carbon market which exploits the unequal terms of trade between core and periphery, which results in the wide chasm in power differential between buyer and seller – or rather, between buyer and seller on the one hand and the intermediaries on the other. If biogas should continue to be supported by carbon trading mechanisms, a number of concerns need to be addressed to make the benefits more equitable

Figure 4.5 A household biogas plant with a *karesa bari* in the background, Godavari, August 2013

Figure 4.6 The biogas fuel is transported to a kitchen in this house, Godavari, August 2013

and to challenge hierarchical and unequal relationships at multiple scales. Key among these are questions about what constitutes a fair trade, who determines the price of commodified carbon, and whether fully informed consent is received for such trade. Meanwhile, more effective and innovative ways to mitigate climate change even in the Nepalese context should be supported. But while efforts to reform the existing process are important, it is equally important to examine the structural drivers that constrain the sustainability and equitability of this approach to development and climate mitigation.

Figure 4.7 The biogas stove (far left) sitting next to an LPG stove, Godavari, August 2013

Towards a multi-scalar political ecology of cow dung

Scholars critical of neoliberal environmental governance question the very basis of carbon trading as inimical to the multiple suggestions for reform just discussed (Barnhart 2013; Bond 2012; Ervine 2018). Placing biogas in the context of the broader issue of climate change and the climate justice discourse, we recognize that it is a modernist and neoliberal solution to the climate crisis, meaning that it is predicated on the commodification of the atmospheric commons, through the buying and selling of entitlements to the atmosphere as a sink, measured in CERs that can be bought and sold in the global market. This approach to climate mitigation and sustainable development can be understood as an extension of the centuries-long project of neoliberalization of nature and environmental governance (Bakker 2005). It follows the logic of a spatial fix, where newer domains are continually sought in order to sustain the system of capitalism (Harvey 2018). As a commodity, the price of carbon credit does not reflect its true ecological value, but rather by the vagaries of the marketplace such that it is greatly underpriced; thus, the goal of internalizing externalities remains unmet (Joshi Forthcoming). Core-periphery dynamics render the buyers of carbon credit – consumers in the Global North, represented by the WB – more powerful in determining this price. They have a vested interest in keeping the price as low as possible, which happens at the cost of ecological degradation and the devaluation of labor (Sheppard et al. 2009). In a buyer's market, the seller of carbon credit – the biogas user/owner – has little to no way in determining this price. Moreover, as we saw, they might not have had a say in whether they wanted to sell their carbon credits. Such unequal terms of trade have led to concerns that the CDM might serve as a neocolonial instrument (Bumpus and Liverman 2010). Rather than facilitate a payment of climate debt, the CDM thus perpetuates North/South inequities, thus adding a fourth dimension to what Roberts and Parks (2007) describe as the triple injustices of climate change – in addition to responsibility, vulnerability, and capacity to address the problem, even solutions devised to address the climate crisis can be inequitable and unjust.

Questioning the legitimacy of CDM biogas projects is not about being anti-development. It is about asking if the claimed benefits are truly accrued, if they are sufficient, and asking who is profiting off of claimed beneficiaries. The distributive and participatory inequities exist not only in a North/South context but at other scales as well, including within Global South and within nations. A politics of scale is at play in the way that the carbon trade is orchestrated: hundreds of thousands of individual households enable profits and decision-making at the site of transaction between the AEPC and the WB. The sellers and presumably the buyers of said carbon credit, who are the ones investing their labor and resources, are left out of the equation, such that the ones who are being exploited and the ones who are made complicit may not be aware of these machinations. Pre-existing social power dynamics, particularly at the site of the household and at the state-society nexus, which also serves as an urban-rural interface, enables the exploitative dimensions of the transaction. Most egregious is the 'accumulation by dispossession' (Harvey 2018) of both the agency of the (often) male head of household, who signs the contract and gives up his ownership, possibly unwittingly – and any future claims – over carbon credit sold for a pittance, and that of his spouse, whose labor enables this entire enterprise without having any say over how it should function, even as her own activities in the kitchen are subject to monitoring and regulation, lest there should be emission leakages. Husband and wife are implicated in a global carbon trade without their fully informed consent or participation in the negotiation of the price of the product they are selling. At successive sites/scales from household to nation to North/South, one party benefits at the cost of the other, and the hierarchical relationships at each site enable the others, illustrating Cox's (1993) notion that there is an inherent relationship between power dynamics in international relations and within nation-states. As Barnhart (2013, 3) articulates:

> The Global North keeps emitting, with minimal lifestyle change, and the burden of behaviour changes to save the planet is displaced to the Global South. . . . What was intended to be common but differentially shared responsibility for reducing emissions has instead become another form of domination by the economically and politically powerful.

Sustainable development in this context occurs within a paradigm of ecological modernization and neoliberal governance that creates governable gendered subjects stripped of their agency to shape policies that affect their lives. While some material benefits exist for household members, in the form of cleaner indoor air quality and money saved from subsidies, they are limited and limiting, as they are encouraged to be ecological consumers rather than ecological citizens, because of the narrowly articulated conceptualizations of sustainable development (Joshi Forthcoming). Neoliberalization alters state-society relations, diminishing the role of the state, local institutions, and weakening social networks in favor of market forces to prevail over livelihood prospects (Bakker 2005). In the ensuing transitions, community self-determination over local resources is often weakened while low-wage jobs in a market-based economy proliferate. These transformations have indeed occurred in Nepal over the time period when biogas as a development project proliferated from the start until its recent reincarnation as a neoliberal climate solution (Barnhart 2013). In this transition, rural people who subsisted on forest resources would increasingly find their access to the forest commons and all the social dynamics associated with it curtailed. Meanwhile, the burning of firewood as a non-renewable resource would be problematized as a contributor to GHG emissions, and modern renewable energy technology would be offered at subsidized cost to facilitate reproductive labor, while enabling the country to capture a greater share of financial resources in a structure that has benefited India and China, its neighbors, more. This approach to climate mitigation and sustainable development appears reductionist, focused on one or two measurable or quantifiable indicators that can be box-checked, as opposed to a more holistic approach that emphasizes the generation of sustainable livelihoods encompassing a more harmonious nature-society relationship that existed prior to the introduction of modernist development and economic globalization. Agencies looking to integrate meaningful sustainable development with climate mitigation goals would do well to heed this insight from Sheppard et al. (2009, 150):

> For people in the third world who derive their livelihoods from the forests, fields, and waters around them, sustainability is intimately related to rights of communal ownership, collectively shared ways of knowing, cultural autonomy, religious rituals, and freedom from externally imposed programs that seek to promote someone else's vision of how to conserve or develop the environments they depend upon.

Reconciling sustainable development and environmental justice

The juxtaposition of the two development discourses in this chapter illustrates how a politics of scale is exercised by multiple sets of actors for different reasons within the sphere of global climate governance. The outcome appears to be to conjure up the material realities of stark disparities between high and low carbon emitters across the globe to make very different claims about how to address the climate crisis. I outline two specific claims pertaining to environmental/climate justice here. One claim pertains to North/South differences in responsibility, vulnerability, and capability, actively made by Indian climate and other officials, drawing on historical economic injustice precipitated by colonial power dynamics between North and South, as well as the low per capita emissions of its citizens, with the goal of arguing against a binding cap on its emissions, as this was seen as antithetical to meeting the basic needs of these citizens. Consequently, EJ here is a 'right to development'. The second claim is not explicitly made as an environmental or climate justice claim but is made by Nepalese climate and clean energy officials about how a climate solution from which it benefits financially benefits the subaltern. An implicit politics of scale here involves drawing on the emission-generating practices of rural households, collating their entitlements to the atmospheric sink, and selling them for economic benefits. In both cases, the emission-generating practices of some of the lowest emission-generating populations on earth are conjured up to benefit the state, in two Global South countries. In both cases, claims are made about a 'right to development' or the 'meeting of sustainable development goals'. Climate justice moves from a 'polluter pays' idea in the North/South context to a 'polluter earns' in the CDM context, albeit with the problematic designation of 'polluter' to the firewood burning household. Arguably, one might say that one state seeks to hide behind the poor; the other seeks to profit from them. Here I find it appropriate to ask these questions: what rights to development does the biogas-adopting household have, are these rights met by the CDM, and how would a North/South climate justice argument help them attain those rights? If a 'right to development' claim similar to one made by Indian climate officials were applied to the Nepalese households – and for Indian households in similar contexts, for that matter – it would translate to a right to produce GHG emissions in a climate of ever scarce atmospheric space. It is paradoxical, then, that the carbon trading mechanism of the CDM, in fact, further encroaches upon those rights. Those rights are, in fact, taken away in exchange for a one-time nominal subsidy for a technological benefit that they are also paying for, more than what they receive as subsidy.

In thinking about climate change through a lens of EJ, we see a fourth dimension of injustice at play: in addition to the triple injustice of responsibility, vulnerability, and capability, there are unjust climate solutions that either exacerbate existing inequities or create new ones. These inequities are not unconnected to the triple injustices. Those that are least responsible also tend to be most vulnerable to not only climate impacts but also the impacts of climate solutions and have the least capability in confronting them, in part because of the distributive and recognition-based inequities. The injustice of climate solutions includes distributive disparities, participation hurdles, as well as absence of recognition – or malrecognition – and this injustice is structurally created by neoliberal capitalism, core-periphery structural inequalities experienced as a legacy of colonialism, patriarchal structures, and a rural-urban divide. A key aspect of the unjust climate solution examined here is the way in which the 'target of emission reduction' is identified and framed as a 'sustainable development beneficiary': a household where firewood is predominantly used to cook, and that also has at least one cow or buffalo, to satisfy the burden of additionality. If you meet this criteria, you can simultaneously serve as a victim to be rescued, a resource-holder to be dispossessed, and a problem-solver of a grave global crisis. As a subaltern member of the Global South, you are not 'othered' in this manner by Global North entities directly but by other, more privileged members in the Global South: urban Western-educated dwellers of your own country. Such targeting is made possible through a phenomenon described as green Orientalism that posits certain ecologically friendly lifestyles suitable for an 'othered' social group (Lohmann 1993). It is also made possible by your governability as a subject of MRV, deemed important in CDM rationalities, as well as your lack of formal education and knowledge of worldly affairs such as carbon trading.

I argue that we should think of the structural explanations of distributive, participatory, and recognition-based understandings of EJ not only in the sense of Marxist political economy of capitalism broadly construed, as has been argued by Low and Gleeson (2002), among others, but also in terms of the impact of the enduring structures of racism, colonialism, and patriarchy, as well as more specific structures such as core-periphery relations. The importance of Indigenous sovereignty and self-determination in settler colonial contexts is a well-established focus of EJ studies (Sze 2020; Whyte 2018), but we can extend this to a global context to think of post- and neocolonial dynamics as an environmental injustice. In the Global South context, such analyses enable us to build more explicit and clear linkages between EJ and sustainable development, particularly in a context where EJ is expressed as a right to development. In doing so, we also have opportunities to build on work that seeks to understand EJ as a struggle for maximizing individual, community, and national capabilities. If EJ is about enhancing capabilities, then sustainable development must be also understood as a way to enhance capabilities. Rather than understanding sustainable development in reductionist ways by checking a number of boxes, we need to think about sustainable development in holistic ways that strive to enhance the capabilities of people to have more control over their lives and livelihoods and, in that pursuit, be empowered to strive to undo the effects of the enclosure and privatization of their local commons, which is by far the most enduring legacy of colonialism (Sheppard et al. 2009).

References

Agarwal, Anil et al. 1999. "Preface." In *Green Politics*, edited by Anil Agrawal, Sunita Narain, and Anju Sharma. New Delhi: Centre for Science and Environment.

Agarwal, Arun, and Sunita Narain. 1990. *Global Warming in an Unequal World*. New Delhi: Centre for Science and Environment.

Anand, Ruchi. 2004. *International Environmental Justice*. Burlington: Ashgate.

Ananthapadmanabhan, G., K. Srinivas, and Vinuta Gopal. 2007. *Hiding Behind the Poor*. Bangalore: Greenpeace India Society.

Baer, Paul, Tom Athanasiou, Sivan Kartha, and Eric Kemp-Benedict. 2009. "Greenhouse Development Rights: A Proposal for a Fair Global Climate Treaty." *Ethics, Policy & Environment* 12, no. 3: 267–81.

Bakker, Karen. 2005. "Neoliberalizing Nature? Market Environmentalism in Water Supply in England and Wales." *Annals of the Association of American Geographers* 95, no. 3: 542–65.

Barnhart, Shaunna. 2013. "From Household Decisions to Global Networks: Biogas and the Allure of Carbon Trading in Nepal." *The Professional Geographer* 66, no. 3: 345–53.

Baxter, Brian. 2005. *A Theory of Ecological Justice*. New York: Routledge.

Bhagwati, Jagdish N., ed. 1977. *The New International Economic Order*. Cambridge, MA: The MIT Press.

Bond, Patrick. 2012. "Emissions Trading, New Enclosures and Eco-Social Contestation." *Antipode* 44, no. 3: 684–701.

Bookchin, Murray. 1990. *Remaking Society*. Boston: South End Press.

Brundtland, Gro H. 1987. *Our Common Future*. Report of the World Commission on Environment and Development. Oslo: United Nations, March 20.

Buckingham, Susan, and Rakibe Kulcur. 2009. "Gendered Geographies of Environmental Justice." *Antipode* 41, no. 4: 659–83.

Bullard, Robert D. 1990. *Dumping in Dixie*. Boulder: Westview Press.

Bumpus, Adam, and Diana Liverman. 2010. "Carbon Colonialism? Offsets, Greenhouse Gas Reductions and Sustainable Development." In *Global Political Ecology*, edited by Richard Peet, Paul Robbins, and Michael Watts. London: Routledge.

CDM Watch. 20102. "Hydro Power Projects in the CDM." Policy Brief, February. https://carbonmarketwatch.org/wp-content/uploads/2009/07/120228_Hydro-Power-Brief_LR_WEB.pdf.

Chiro, Giovanni Di. 2008. "Living Environmentalisms: Coalition Politics, Social Reproduction, and Environmental Justice." *Environmental Politics* 17, no. 2: 276–98.

Cole, Luke W., and Sheila R. Foster. 2000. *From the Ground Up*. New York: New York University Press.

Cox, Robert W. 1993. "Gramsci, Hegemony and International Relations: An Essay in Method." In *Gramsci, Historical Materialism and International Relations*, edited by Stephen Gill. Cambridge: Cambridge University Press.

Crutzen, Paul. 2002. "Geology of Mankind." *Nature* 415, no. 23.

Dhakal, Shobhakar, and Anil Raut. 2010. "Potential and Bottlenecks of the Carbon Market: The Cases of a Developing Country, Nepal." *Energy Policy* 38, no. 7: 3781–89.

Dobson, Andrew. 1999. *Fairness and Futurity*. New York: Oxford University Press.

Dreze, Jean, and Amartya Sen. 2002. *India*, 2nd ed. Oxford: Oxford University Press.

Ervine, Kate. 2018. *Carbon*. Cambridge: Polity Press.

Finley-Brook, Mary, and Curtis Thomas. 2011. "Renewable Energy and Human Rights Violations: Illustrative Cases from Indigenous Territories in Panama." *Annals of the Association of American Geographers* 101, no. 4: 863–72.

Gilio-Whitaker, Dina. 2019. *As Long as Grass Grows*. Boston: Beacon Press.

Government of India. 2008. "National Action Plan on Climate Change: Prime Minister's Council on Climate Change." http://pmindia.gov.in/climate_change_english.pdf.

Government of Nepal. 2019. "Progress at a Glance: A Year in Review, Fiscal Year 2075–76 (2018–19)." Alternative Energy Promotion Center, Ministry of Energy, Water Resources and Irrigation. https://www.aepc.gov.np/documents/annual-progress-report-aepc.

Griffin, Paul. 2017. "The Carbon Majors Database." July. https://b8f65cb373b1b7b15feb-c70d8ead6ced550b4d987d7c03fcdd1d.ssl.cf3.rackcdn.com/cms/reports/documents/000/002/327/original/Carbon-Majors-Report-2017.pdf.

Hansen, Roger D. 1980. "North/South Policy-What's the Problem?" *Foreign Affairs* 58, no. 5: 1104–28.

Hardin, Garrett. 1968. "The Tragedy of the Commons." *Science* 162: 1243–48.

Harvey, David. 1996. *Justice, Nature and the Geography of Difference*. Cambridge: Blackwell.

———. 2018. *Marx, Capital and the Madness of Economic Reason*. New York: Oxford University Press.

India. n.d. "INDC: India's Intended Nationally Determined Contribution: Working Towards Climate Justice." www.indiaenvironmentportal.org.in/content/419700/indias-intended-nationally-determined-contribution-working-towards-climate-justice/.

IPCC. 2014. "Climate Change 2014: Synthesis Report." In *Fifth Assessment Report of the Intergovernmental Panel on Climate Change*, edited by Core Writing Team, Rajendra K. Pachauri, and Leo A. Meyer. Geneva: IPCC.

Joshi, Shangrila. 2014. "Environmental Justice Discourses in Indian Climate Politics." *Geo-Journal* 79, no. 6: 677–91.

———. 2015. "Postcoloniality and the North/South Binary Revisited: The Case of India's Climate Politics." In *The International Handbook of Political Ecology*, edited by Raymond L. Bryant. Cheltenham: Edward Elgar.

———. Forthcoming. "A Political Ecology of Neoliberal Climate Mitigation Policy in Nepal." Conference Proceedings 2019: The Annual Kathmandu Conference on Nepal and the Himalaya, Kathmandu.

Kurtz, Hilda E. 2009. "Acknowledging the Racial State: An Agenda for Environmental Justice Research." *Antipode* 41, no. 4: 684–704.

Laclau, Ernesto. 1990. *New Reflections on the Revolution of Our Time*. New York: Verso.

———. 2000. "Constructing Universality." In *Contingency, Hegemony, Universality*, edited by Judith Butler, Ernesto Laclau, and Slavoj Žižek. London: Verso.

Lele, Sharachchandra M. 1991. "Sustainable Development: A Critical Review." *World Development* 19, no. 6: 607–21.

Lohmann, Larry. 1993. "Green Orientalism." *Ecologist* 23, no. 6: 202–4.

Low, Nicholas, and Brendan Gleeson. 2002. "Expanding the Discourse: Ecosocialization and Environmental Justice." In *Environmental Justice*, edited by John Byrne, Leigh Glover, and Cecilia Martinez. New Brunswick: Transaction Publishers.

Maniates, Michael. 2001. "Individualization: Plant a Tree, Buy a Bike, Save the World?" *Global Environmental Politics* 1, no. 3: 31–52.

Marston, Sallie A., John P. Jones, and Keith Woodward. 2005. "Human Geography Without Scale." *Transactions of the Institute of British Geographers* 30, no. 4: 416–32.

Meadows, Donella H., Dennis L. Meadows, Jorden Randers, and William W. Behrens. 1972. *The Limits to Growth*. New York: Universe Books.

Mohai, Paul, David Pellow, and J. Timmons Roberts. 2009. "Environmental Justice." *The Annual Review of Environment and Resources* 34: 405–30.

Narain, Sunita. 2019. "Equity: The Final Frontier for an Effective Climate Change Agreement." In *Climate Futures*, edited by Kum-Kum Bhavnani, John Foran, Priya A. Kurian, and Debashish Munshi. London: Zed Books.

NESS. 2011. *Final Report on: Annual Biogas Users' Survey 2009/10 for Biogas Support Program-Nepal (BSP-Nepal) Activity II*. Kathmandu: Nepal Environmental and Scientific Services, Ministry of Environment, Alternative Energy Promotion Center, Government of Nepal, July.

Neumann, Roderick P. 2009. "Political Ecology: Theorizing Scale." *Progress in Human Geography* 33, no. 3: 398–406.

Ostrom, Elinor. 2010. "Polycentric Systems for Coping with Collective Action and Global Environmental Change." *Global Environmental Change* 20, no. 4: 550–57.

PDD. 2017. "Gold Standard for the Global Goals." Key Project Information & VPA Design Document, July. https://www.atmosfair.de/wp-content/uploads/1-gs4gg_vpa-dd_cpa-10.pdf.

Pulido, Laura. 2016. "Flint, Environmental Racism, and Racial Capitalism." *Capitalism Nature Socialism* 27, no. 3: 1–16.

Rangan, Haripriya, and Christian A. Kull. 2009. "What Makes Ecology 'Political'?: Rethinking 'Scale' in Political Ecology." *Progress in Human Geography* 33, no. 1: 28–45.

Redclift, Michael. 2005. "Sustainable Development (1987–2005): An Oxymoron Comes of Age." *Sustainable Development* 13: 212–27.

Roberts, J. Timmons. 2007. "Globalizing Environmental Justice." In *Environmental Justice and Environmentalism*, edited by Ronald Sandler and Phaedra C. Pezzullo. Cambridge, MA: The MIT Press.

Roberts, J. Timmons, and B. C. Parks. 2007. *A Climate of Injustice*. Cambridge, MA: The MIT Press.

Roos, Bonnie, and Alex Hunt. 2010. *Postcolonial Green*. Charlottesville: University of Virginia Press.

Schlosberg, David. 2009. *Defining Environmental Justice*. New York: Oxford University Press.

Sen, Amartya. 1999. *Development as Freedom*. New York: Anchor Books.

Sheppard, Eric, Philip Porter, David Faust, and Richa Nagar. 2009. *A World of Difference*. New York: The Guilford Press.

Smith, Neil. 1993. "Homeless/Global: Scaling Places." In *Mapping the Futures*, edited by Jon Bird et al. New York: Routledge.

Sultana, Farhana. 2013. "Gendering Climate Change: Geographical Insights." *Professional Geographer* 66, no. 3: 372–81.

Sundberg, Juanita. 2011. "Diabolic Caminos in the Desert and Cat Fights on the Rio: A Posthumanist Political Ecology of Boundary Enforcement in the United States-Mexico Borderlands." *Annals of the Association of American Geographers* 101, no. 2: 318–36.

Swyngedouw, Erik, and Nikolas C. Heynen. 2004. "Urban Political Ecology, Justice and the Politics of Scale." *Antipode* 35, no. 5: 898–918.

Sze, Julie. 2020. *Environmental Justice in a Moment of Danger*. Oakland: University of California Press.

Taylor, Dorceta E. 2000. "The Rise of the Environmental Justice Paradigm: Injustice Framing and the Social Construction of Environmental Discourses." *American Behavioral Scientist* 43, no. 4: 508–80.

UNEP/DTU. 2020. *CDM Projects by Host Region*. Copenhagen: Centre on Energy, Climate and Sustainable Development, September 1. www.cdmpipeline.org/.

UNFCCC. 2020. "Clean Development Mechanism." September 16. https://cdm.unfccc.int/.

Walker, Gordon. 2009. "Beyond Distribution and Proximity: Exploring the Multiple Spatialities of Environmental Justice." *Antipode* 41, no. 4: 614–36.

———. 2012. *Environmental Justice*. New York: Routledge.

Whyte, Kyle P. 2018. "Settler Colonialism, Ecology, and Environmental Justice." *Environment & Society* 9, no. 1: 125–44.

Williams, Glyn, and Emma Mawdsley. 2006. "Postcolonial Environmental Justice: Government and Governance in India." *Geoforum* 37: 660–70.

5 From forest commons to carbon commodities – REDD+ and community forestry in Nepal

Forests are the lungs of the earth, because they absorb carbon dioxide and give off life-giving oxygen. Beyond this basic bio-chemical and ecological function, forests have social, cultural, spiritual, religious, and economic significance; they are valued for one or more of these attributes by different groups. In the contemporary Western world, forests are often associated with environmentalism, indicated by the term 'tree-hugger' for environmental activists, as the Himalayan women representing the *Chipko* movement were thought to be, although what the *Chipko* women were fighting for is open to debate (Rangan 2000). In more ancient mythologies and literature – such as in the Sanskrit epic *Ramayan* penned by *Adikavi* Valmiki in the forests of Chitwan, Nepal – forests were a place of learning, exile, and sacrifice for the ruling class; but also a place for meditation, hunting and gathering, and making a livelihood for commoners. Today forests continue to mean different things to different groups of people. Some people continue to live in them, or live nearby while they rely on forest resources for their livelihood; others romanticize them from a distance, fight for certain forest imaginaries; yet others may visit occasionally for recreation or research. Since knowledge of a global ecological crisis came into popular consciousness in the 1960s, forests have been regarded a global heritage and resource – a global commons – with a consequent set of institutions and programs designed to address deforestation. The economic implications of forest-based resources – not least of which is timber – have rendered forests a key aspect of national resource policy worldwide. Going back at least to the era of European colonization in South Asia and elsewhere, forests – as a home for productive and exploitable natural resources – have functioned as a space/site of control and contestation, capitalist accumulation by dispossession, governmentality, as well as community and cooperation (Agrawal 2005; Damodaran 2006; Ostrom 1999a, 1999b). Today, forests continue to be a site of contestation, where identity, livelihoods, and ecological stability are all at stake in the ways in which contestations over the right way to configure forests unfold in particular places.

It is in this variegated context that we should seek to understand the rapidly unfurling role of forests in addressing the climate crisis. Due to the carbon sequestration potential of forests, their protection has been increasingly viewed as a way to mitigate climate change. Moreover, due to the livelihood prospects of forests, as well as the protections they are known to provide from storm surges, they are increasingly regarded as important for climate adaptation and resilience as well (Corbera and Schroeder 2017; MoFE 2019a). While the benefits of forests and their global to local significance are undisputed, from a political ecology perspective, it matters who is leading and shaping the discourse and policies around forest-climate linkages, and for whom; how these discourses and policies are being contested, to what effect, at particular sites and scales of governance; how new forms of forest governance build off the old, and how the resultant new configurations are affecting the lives of the forest-dependent communities.

Reducing Emissions from Deforestation and Forest Degradation (REDD)

Although negotiations to formulate a global climate treaty started in 1992, it was not until a decade later that forests would come to play a role in international deliberations over global climate governance. The United States and other countries of the Global North had strenuously pushed for a forest convention during the Rio Earth Summit, but faced with staunch positions from countries such as Malaysia, Brazil, and India, who collectively saw a global forest treaty as motivated by an economic agenda to benefit the North, as well as a potential threat to national and local sovereignty in the South, it would not come to pass (Agarwal and Narain 1990; Gulbrandsen 2012). The initial impetus for REDD was a proposal for 'compensated reduction' for preventing deforestation introduced during COP-9 in 2003 by Brazil; Papua New Guinea and Costa Rica submitted country position papers on REDD in 2005; REDD was adopted as the new forest and climate change regime to succeed the Kyoto Protocol, as per the Bali Road Map during COP-13 in 2007, aiming to "compensate developing countries and their communities for their forest conservation and regeneration efforts" (Poudel, Rana, and Thwaites 2018, 148). The 2007 Bali Action Plan "contained the building blocks of what has ultimately become the 2015 Paris Agreement" (Corbera and Schroeder 2017, 1).

During COP-14 in Poznan in 2008, REDD would be reframed as REDD+ to encompass a broader meaning: reducing emissions from deforestation and forest degradation and the conservation, sustainable management of forests, and enhancement of forest carbon stocks (ibid). This was when piloting initiated. Following the Cancun Agreement in 2010, the idea of environmental and social safeguards was introduced, with an articulation of Indigenous rights. When the Paris Agreement was adopted in 2015 by 197 countries, REDD+ was institutionalized to facilitate climate mitigation, adaptation, and finance from 2020 and onward under the aegis of the UNFCCC (Dhungana, Poudel, and Bhandari 2018; Poudel, Rana, and Thwaites 2018; Rai, Pant, and Nepal 2018; United Nations 2015). Intriguingly, the Paris Agreement makes no references to carbon trading or offsets. Rather, it uses the language of 'internationally transferred mitigation outcomes' and 'results-based payments'. Article Five of the Paris Agreement states:

> Parties are encouraged to take action to implement and support, including through results-based payments, the existing framework as set out in related guidance and decisions already agreed under the Convention for: policy approaches and positive incentives for activities relating to reducing emissions from deforestation and forest degradation, and the role of conservation, sustainable management of forests and enhancement of forest carbon stocks in developing countries; and alternative policy approaches, such as joint mitigation and adaptation approaches for the integral and sustainable management of forests, while reaffirming the importance of incentivizing, as appropriate, non-carbon benefits associated with such approaches.
>
> (United Nations 2015, 6)

Results-based finance (RBF) is a new iteration of development finance that aims to deliver financial rewards after desired results have been demonstrated, which in the case of REDD+ would be reduction of GHG emissions from carbon sequestration in forests, measured in metric tons of CO2 equivalent (MtCO2) (Stumpf, Kleymann, and Windhorst 2018). It is meant to incentivize countries into enacting long-term transformational changes in land use and rural development so as to scale up and normalize low-deforestation scenarios into the future. Countries seeking RBF need to meet certain 'conditionalities', including the creation of a national strategy or action plan and the establishment of mechanisms for accounting, monitoring, and reporting of the results desired, which includes social and environmental safeguards (ibid). REDD+ implementation is planned over three overlapping phases. The first two phases are funded with bilateral and multilateral development support: a readiness phase, where countries focus on developing their institutional capacities through policy and legislation, and an implementation phase. The third phase is when the financing occurs, through payments from various sources for verified emission reductions. In the 2013 Warsaw conference, it was determined that the Green Climate Fund would finance REDD+ as part of an international climate finance regime. While this has yet to be materialized, RBF has served as 'bridge financing', primarily from public sources, such as the governments of Germany, Norway, and the UK (ibid). Forest Carbon Partnership Facility's (FCPF) Carbon Fund managed by the WB has been a prominent entity to pilot RBF for REDD+. There is a formal application process where countries submit an 'ER Programme Idea' followed by the Emission Reduction Purchase Document (ERPD) for emission reduction. If accepted, these countries enter an agreement to be compensated for documented GHG reductions after verification. REDD+ started to move from preparedness to early implementation, with some countries – including Nepal and Vietnam in Asia – signing Emission Reduction Purchase Agreements (ERPAs) in 2018–9.

Once REDD+ pilots started to take off, they met with staunch objections from environmental and Indigenous rights groups based on multiple concerns: incentivizing governments to disenfranchise forest-dependent communities through land grabs, letting polluting companies claiming carbon neutrality too easily off the hook, enabling shrewd parties to overinflate baseline scenarios to earn more offsets, and overall for promoting false solutions rather than preventing the burning of fossil fuels (Ervine 2018; Gilbertson 2017). In a keynote address given at the Indigenous Climate Justice Symposium at The Evergreen State College in 2015, Indigenous Environmental Network (IEN) Executive Director Tom Goldtooth conveyed a clear message about his organization's position: "No REDD+". During the 2019 COP-25 held in Spain, where Article Six of the Paris Agreement had been a key point of contention, he has said that carbon markets "will only deepen the climate crisis and shift the burden of carbon trading to the Global South. Indigenous people's territories, rights, and self-determination are all threatened by carbon offsets and the privatization of Mother Earth" (Carraher-Kang 2019). IEN, in collaboration with the Climate Justice Alliance, has published a critical manual on carbon markets that states:

> Carbon trading, carbon offsets and REDD+ are fraudulent climate mitigation mechanisms that in fact help corporations and governments keep extracting and burning fossil fuels. Token revenues distributed to environmental justice communities from carbon trading or carbon pricing can never compensate for the destruction wrought by the extraction and pollution that is the source of the revenue. Accepting such revenue not only does not compensate for the damage to our air, bodies, environment, and nature, but also implicates the receiver in the extraction, pollution and natural disasters that such pollution causes.
>
> (Gilbertson 2017, 4–5)

Indeed, Corbera and Schroeder (2017, 1) note that REDD+ was initially embraced as a 'cheap' way to mitigate climate change that centered on "the idea of payment for standing forests". Operationalization challenges would later reveal that if the social safeguards were taken seriously, it would not be so cheap after all. Since REDD+ pilot projects started to unfold in 2008, academic research on the subject has burgeoned, with the body of work reflecting both skepticism and optimism, with even the most optimistic studies demonstrating challenges to the successful implementation of REDD+. Key questions about its legitimacy involve top-down decision-making tendencies and "the problematic absence of social actors that play a key role in development planning and land-use change" (ibid, 3). On the other hand, REDD+ contributed to improving land use and forest governance in some contexts, particularly where successful forest governance reforms had created conducive environments for better outcomes (ibid). Kane et al. (2018) note that REDD+ could exacerbate or facilitate the resolution of conflicts in forested landscapes, depending on how it was operationalized. Its transformative potential for empowering tradition-ally marginalized groups through representation and tenure has been observed, particularly in relation to its emphasis on the 2010 Cancun Safeguards. This includes provisions for Free, Prior and Informed Consent (FPIC), particularly for Indigenous people, in adherence to the 1989 Convention on the rights of Indigenous and Tribal Peoples (ILO Convention 169), and the 2007 UN Declaration of the Rights of Indigenous People, which have been emphasized in order to enhance REDD+ legitimacy, although their implementation has so far left much to be desired (Satyal 2018).

Advocates of REDD+ emphasize environmental and social co-benefits beyond carbon sequestration, including livelihood enhancement, healthier forests and biodiversity conservation, improved governance, land tenure, and inclusion in decision-making processes – benefits that are not a given but rather may have to be fought over, given the possibility that the opposite may result (Poudel, Rana, and Thwaites 2018). Given claims that this relatively new international policy instrument could yield beneficial outcomes for local communities under the right circumstances, and in contexts where robust social structures are already in place, a critical examination of the REDD+ experience in Nepal is warranted. This is a country where one of the earliest, most successful, and enduring experiments with community-based forest management (CBFM) have taken place (Thwaites, Fisher, and Poudel 2018). Community-based forestry has many variants, but the key idea is the role played by local people in governing forest resource. The strongest of these variants is 'community forestry with full devolution'. Thwaites, Fisher, and Poudel (2018, 5–6) assert that community forestry as it is practiced in Nepal is

> one of the more radical programs in terms of the transfer of rights to the local community. . . . Nepal could be described as a living laboratory for the implementation of community forestry and testing of theories of democratization and community-based NRM. In this context, it is rel-evant to take a critical look at what is happening, good and bad, and how it is changing in Nepal, and how that might inform people's thoughts in Nepal and elsewhere.

An examination of how REDD+ might transform the governance of forests in Nepal provides insight as to the consequences of the commodification of the global atmosphere for the localized forest commons.

Community forestry in Nepal

The Nepalese variant of community forestry (CF) – one of several forms of CBFM practices – is regarded as a model to emulate (Thwaites, Fisher, and Poudel 2018). More than 40 percent of Nepal's territory is forested, with a third of Nepal's forests designated as 'community forest' (about 2.2 million hectares out of 6.6 million hectares), managed by 22,645 community forestry user groups (CFUGs) with the involvement of more than three million households as of June 2020, compromising almost half of Nepal's population; if other CBFM groups are included, there are more than 30,000 forest groups (email correspondence with MoFE official, October 2020; MoFE 2018, 2019a). Since the formal introduction of CF, over 1.2 million hectares of degraded forests had been restored by 2015 (Manandar and Parker 2018). Beyond facilitating forest recovery, CF in Nepal has assumed the form of an extensive political, economic, and social system and is considered an effective vehicle for rural development, leadership building, and "a strong political force for democratic change" (Pokharel et al. 2008, 87). Indeed, CF offers a real-world model of a dynamic and productive socio-natural system where ordinary people are in charge of – if not owning – the means of production, albeit under evolving constraints. As well, the case of CF illustrates a process of EJ in a localized Global South context, where distributive, participatory, and recognition-based claims to justice are pursued in regard to forest-based communities (Satyal 2018). I suggest that these progressive outcomes have been made possible through structural transformations in how the nation-state pursues forest governance.

Community forestry in Nepal is often traced to the 1970s as an 'environment and development' agenda supported by international donors, but the idea of community forests can be traced back to a 1953 draft forest policy document (Pokharel et al. 2008). Its use in traditional forest governance practices such as *kipat* – an ancient form of community land tenure practiced in Nepal – predates any official forest policy (Bartlett and Malla 1992). The community forestry intervention in the Nepalese context was a concerted effort to institutionalize Indigenous resource governance strategies already practiced in the country to thwart perceived widespread forest degradation (Gilmour 1989; Ojha et al. 2008). The first structural reforms to facilitate decentralized forest governance in Nepal can be traced to 1978, during the *Panchayat* era; the 1993 Forest Act congealed the CFUG as a political entity, and while it has gone through various transmutations, it remains the foundation of CF as we know it in Nepal today. Ojha et al. (2008, 37–38) assert that forest governance has shifted course in the history of Nepal in response to a number of factors: "local political economy of land-based resources, national political changes, global environmental discourses, and the rhetoric of international development agencies. The present-day practices and policies of community-based management are thus a result of a complex field of influence by these forces". Even as community forestry has been regarded as an exceptional case where 'accumulation without dispossession' (Paudel 2016) has been made possible by these structural reforms, there are concerns that with new interventions, such as REDD+, there is a reasonable fear that the gains of CF may be reversed with increasing pressures towards neoliberalization (Joshi Forthcoming). The politics of CF have been fraught from the start, and these dynamics have not changed as it evolves and comes under new pressures and constraints. The contested nature of CF politics is, I believe, a sign of its vitality and resilience.

Situating community forestry in Nepal's history of land tenure

The development of CF and subsequent reforms in Nepal need to be situated in the context of what was happening globally and its reverberations here. Following the emergence of an 'environmental movement' in the 1960s in the United States and other First World nations, and the consequent Stockholm Conference on the Environment in 1972, the post–World War II development aid conditionalities were increasingly geared towards an environmental agenda. In particular, donors in Nepal were responding to the then compelling 'Theory of Himalayan Environmental Degradation' (Eckholm 1975) – a theory that has since been debunked as a 'pseudo-crisis' narrative by political ecologists for its neo-Malthusian orientation and exaggerated claims (Blaikie and Brookfield 1987; Gilmour 1989 – by investing in programs aimed at reversing the alarming rates of deforestation perceived at the time from a Western gaze. Not that deforestation was not occurring or that it was occurring for the first time. Timber was central to what we now consider Nepal's artistic and architectural heritage – the name Kathmandu of Nepal's capital city comes from the word *Kashthamandap* (temple of wood, *kath* meaning wood) built during the 7th–12th century, constructed from a single tree as per oral history (Shrestha 2016). But an ecological consciousness about the significance of forests and their protection was present early on, as evidenced by traditional *kipat* systems of forest conservation (Bartlett and Malla 1992), as well as in the *Fourteenth Edict* of King Ram Shah (1606–1636):

> If there are no trees, there will be no water whenever one looks for it. The watering places will become dry. If forests are cut down, there will be avalanches. If there are many avalanches, there will be great accidents. Accidents also destroy the fields. Without forests, the householder's work cannot be accomplished. Therefore, he who cuts down the forest near a watering place will be fined five rupees.
>
> (cited in Bista 1991, x)

During the British colonial era, Nepal's sovereignty remained intact, but its resources would be exploited by the British. Particularly during Nepal's Rana regime (1846–1951), Rana courtiers – having effectively seized power from the Shah monarchy to become de facto rulers for a century – "extensively exploited the Terai forest for [export] to India, primarily for Railway sleepers, to appease the British in India, gain revenue and secure political autonomy from the latter" (Ojha et al. 2008, 38). Northern India's *Sal* forests had been decimated by the turn of the 19th century, and Terai forests served as a spatial fix for the insatiable appetite for timber in British India; this was arguably the first instance of technologically aided extractivism experienced in the Nepalese context – with the first saw mills established in 1900–1 in the Terai – even as people therein had been engaging in regional and trans-Himalayan trade since at least as far back as the 7th century (Harvey 2018; Mulmi 2017; Whelpton 2005).

Forests in the Terai at the border were preserved by the Nepali state as a geophysical barrier against possible territorial encroachment by British India. After India gained Independence from the British in 1947 – and helped the Shah monarchy overcome Rana rule in 1951 – a flood of post–World War II 'developmentalist' intervention from the West entered Nepal in the 1950s (Whelpton 2005). Malaria eradication with DDT in the Terai forests was one of the outcomes, such that those who were not Indigenous to the region and had not developed immunity to malaria – as the Tharu had – could settle there. In 1963, the Shah-ruled Nepali state initiated a resettlement program from the (land-slide-prone and resource-poor) hills and Nepalis abroad (particularly India and Burma) to settle in the fertile, productive, and densely forested lands of the Terai, leading to considerable deforestation and forest degradation throughout the 1970s, as well as displacement of the Indigenous people in the area – Chepang, Tharu, Bote, Majhi among others. While these groups were pushed to the margins, already wealthy landlords, state nobilities, royalists, ex-military servants of the state – mostly of Khas/Arya heritage, who are today's Bahun/Chhetri – claiming to be landless were able to receive forested land as *jagir*, a system of rent extraction to exert political control during the halcyon days of the Monarchy. Much of these forests would be cleared for agriculture and revenues from the sale of timber (Ojha et al. 2008), most of which would be siphoned off to Kathmandu, which had become the core of the Nepali state economy since its subjugation by the Shah rulers of Gorkha in 1768/9, with the Terai being relegated to the periphery (Blaikie and Brookfield 1987; Whelpton 2005). Following the ousting of the Rana regime in 1951 – with the support of a newly independent India – the Private Forest Nationalisation Act 1957 was implemented, leading to a wave of nationalization of the region's forests. This takeover would have been an enclosure of the commons except for the fact that they were mostly privately owned by the ruling gentry. The nationalization of forests was in a way a taking back of the privatized forest commons from the landed gentry favored by the Rana and Shah regimes, at a time when a growing people's movement in the 1950s led to democratization within the Monarchy, with King Mahendra reluctantly ushering in a multiparty elected parliament and a Ministry of Forests in 1959. However, a centralized forest bureaucracy reluctant to share power with local communities – the proposal for community forestry in the 1953 draft forest policy was absent in the 1957 Forest Act – proved inadequate to reverse the tide of deforestation, especially when it criminalized traditional practices and antagonized local people – a 1967 Special Act authorizing forestry officials to "shoot forest offenders at sight" was a case in point (Ojha et al. 2008, 39). As a result of these exercises of state authority unregulated extraction, rampant illegal cross-border logging, and encroachment of forests had taken root.

It was in this political climate that the idea of a participatory forestry program was introduced in Nepal in the 1970s. International donors that had until the mid-1970s supported a top-down preservationist agenda, such as the first protected area, the (Royal) Chitwan National Park in 1973, were now undergoing a paradigm shift towards decentralization and embracing the discourse of 'participatory development' and starting to lose faith in the state's ability to meet their environmental agenda (Ojha et al. 2008; Ojha and Timsina 2008). Nepal at the time was already ensconced in a developmentalist mindset, with state agencies happy to accommodate the agendas of foreign funders (Bista 1991; Whelpton 2005). Donors' emphases on participatory development, good governance, environmental imaginaries tethered to forest vitality, coupled with populist sentiments of democracy – embodied by the People's Movement of 1990 – against the vestiges of Rana rule and the power-centralizing tendencies of the Monarchy led to ideal circumstances for the birth of the CF regime in Nepal. Following a period of development starting with the National Forestry Plan of 1976, followed by the Decentralization Act 1982, the 1987 and 1989 Master Plans for Forestry Sector prepared with international technical and financial assistance the 1993 Forest Act, officially promulgated CF, and initiated the process that would formally recognize CFUGs as autonomous and self-governing organizations (Ojha et al. 2008).

CF is distinct from other forms of CBFM practiced in Nepal by dint of the exclusivity affiliated communities enjoy "by excluding non-members" geographically delimited (ibid, 29) and is often discussed as an example of successful commons governance at the local scale (Ostrom 1999a, 1999b). Of the six forms of CBFM, the highest degree of autonomy is practiced in CF (Ojha and Timsina 2008). There is an elaborate process of CFUG formation involving – as per the 2002 CF Operational Guidelines – socialization and early negotiations between the Division Forest Office (DFO) staff and forest users, CFUG formation and constitution building, Forest Operational Plan preparation and approval, and finally the handing over of forest including devolution of power and authority to CFUGs that are comprised of an 11–19-member Executive Committee that makes everyday decisions on behalf of members at large, who are allowed to harvest resources from the forest as per the rules stipulated by the Committee. In some CFUGs members pay dues that contribute to the operating costs of the CFUG. Members can have rotating duties, such as guarding the forest in pairs against illegal harvests and forest fires. These functions may be assigned to salaried individuals. Every CFUG makes their own rules – within the constraints prescribed by the government – but generally CFUG members are allowed to harvest firewood and grass for household use; a limited volume of timber is sold at subsidized rates (half of the market price) to members for household construction; and if any timber is permitted to be sold outside the community by the DFO, the revenue generated is taxed. Progressively steep taxation has been a cause for much furor among CFUG members in recent years (Mandal 2019, 2020). As the scope and scale of CFUGs have expanded, so has the number of stakeholders within civil society, government sectors, professional associations, as well as continued engagement of community and donors. The major civil society organization representing CFUGs is the Federation of

Community Forestry Users-Nepal (FECOFUN), with membership of more than 70 percent of CFUGs. Community forests in Nepal have not only been largely self-sufficient financially – with 70 percent of their operating budget generated from the sale of forest products, 16 percent of the budget coming from donors, and 13 percent from the government in 2005 – they have been investing in local community and infrastructure development work, easing the burden on the central and local governments. As per the 2014 Community Forestry Guidelines, CFUGs are expected to invest 25 percent of their revenue on forest conservation, 35 percent on poverty eradication and livelihoods, and the remaining 40 percent on social development (Mandal 2020). Following the royal coup of February 2005, Nepal lost donor support for CF and other programs by more than half, but CF has prevailed, demonstrating its sustainability and resilience as an institution (Pokharel et al. 2008).

Community forestry is not perfect

CF was indeed a 'radical reform' precipitated by a number of endogenous and exogenous forces, but it was far from perfect, and it would go on to become a weaker version of its original intent over time. Two layers of power asymmetry have been documented in the operationalization of CF – one between the state and the community, and two, between dominant and marginalized groups in the community (Ojha and Timsina 2008). Firstly, even as CFUGs were empowered as autonomous agents of governance, the land is legally owned by the state. Beyond land tenure, the state retained control over the decision-making and policy-formation processes, often through "technocratic domination", bringing about an environment of "centralized decentralisation", which was further exacerbated through periodic "reversals" (Pokharel et al. 2008). The most recent iteration of this appears to be occurring in the context of preparing the country for REDD+. New provisions to tax CF revenues at the local, provincial, and federal level, in some cases amounting to a 90 percent tax, were seen by many CFUG members interviewed as an unfair and unilateral move. Beyond being unpopular, it wasn't clear why it was introduced and how exactly it would be implemented.

> *Samudayik Ban* [CF] was created to enhance the life chances of the ordinary person, to give them a better life. That is why forests were handed over, but when you look at what is happening today, the DFO exerts a lot of power. Where are the rights for the forest user? They are with the DFO. That is not right. The rights of *Jal Zameen Jangal* should be with the forest dwellers, the *Majhi* [fisherpeople].
>
> (Bahun/Chhetri male, FECOFUN, Pokhara, August 2019)

Secondly, a 'community' is not a monolith; communities are heterogeneous in their needs, interests, experience, and capabilities, and they often exhibit hierarchical power dynamics; hence the autonomy of a 'community' does not automatically lead to equitable outcomes in benefit-sharing and deliberative processes (Ojha et al. 2008). The poorest and most marginalized groups – including the Indigenous people, who happen to live in flood-prone areas near river banks, and therefore more climate vulnerable in several ways – have tended to be left out of meaningful deliberations and economic benefits, with community elite – sometimes Indigenous but often *Bahun/Chhetri* – enjoying more status, influence, and power in CFUG governance – as a result of which other forms of CBFM such as Leasehold Forestry were introduced with the specific goal of attending to the poorest households. Forest governance at multiple scales has also proved to be exclusive and inequitable in terms of gender and caste. As a result it is fair to say that "CFUG committees [are] still frequently a reflection of inherent social power structures and show capture by influential classes in the community. . . . Disproportionate costs of community forest protection are still carried by the poor" (Pokharel et al. 2008, 65).

Reflecting evolving donor agendas, the community forestry apparatus has over the years embraced issues of gender, livelihoods, poverty, caste, and Indigenous rights, beyond the original issues of deforestation, basic needs, governance, democracy, and decentralization (ibid). Despite ongoing reversals in devolution of power from the state to communities, there have been progressive reforms in areas of gender inequity and Indigenous rights. Since 2014, for instance, a requirement for at least 50 percent of CFUG Executive Committee members – including in the two key decision-making positions of chairperson and secretary – to be held by women has been instituted (Gurung and Gurung 2018). Civil society groups such as FECOFUN, Nepal Federation of Indigenous Nationalities (NEFIN), Dalit Alliance for Natural Resources-Nepal (DANAR), and Himalayan Grassroots Women's Natural Resource Management Association (HIMAWANTI) have been created to enhance representation of traditionally marginalized groups. There have been a number of factors incentivizing these efforts. One is a social environment in the country where bottom-up movements such as the People's Movement of 1990 and the Maoist Movement (1996–2006) have forced the hand of the government desirous of becoming a modern nation-state to institute more progressive laws and policies for greater equity and proportional representation for women, Dalit, and Indigenous People or Adibasi/Janajati. The 2015 Constitution addresses these sentiments by recognizing underrepresented groups and by stipulating the need for reservations – similar to affirmative action – in education, health, housing, food sovereignty, and employment. Notably, groups in the Terai, such as the Indigenous Tharus and the Madhesis, have not been adequately represented still. A second impetus for progressive transformations owes to Nepal's decision to ratify the UN Declaration on the Rights of Indigenous People (UNDRIP) and the Indigenous and Tribal Peoples Convention (ILO-169) in 2007, which obligates Nepal to treat the provisions therein as national law. Despite these ratifications and formal gestures, Nepali society in general and community forestry in particular are still enmeshed in inequitable conditions along lines of caste, class, and gender, with problems of techno-bureaucracy, elite capture, and a dominant state still prevalent (Satyal 2018).

New directions for community forests and preparations for REDD+

The 2000s were a time of turbulent transitions in Nepal. A Maoist revolution that started in the 1990s in Western Nepal moved from margin to center and helped precipitate a People's Revolution in 2006 and the subsequent abolishment of the (almost) two-and-a-half-century-old Monarchy. As per the Interim Constitution of 2007, the Federal Democratic Republic of Nepal was created in 2008. International aid in the development and forestry sectors has dwindled significantly. As with biogas, sources of foreign inflows of finance to the state have gradually transitioned to a more transactional basis for financing community forests, even if not carbon trading per se. Consequently, a strong influence of global climate discourse is apparent in the new directions taken by forest governance in Nepal at the turn of the century. In 2008–9 the MoFSC worked with the FCPF and organizations such as Worldwide Fund for Nature (WWF) Nepal, Asia Network for Sustainable Agriculture and Bioresources (ANSAB), FECOFUN, and the International Centre for Integrated Mountain Development (ICIMOD) to initiate a readiness process for REDD+. A REDD+ Readiness Preparation Proposal (RPP) was approved in 2010, and that year Nepal became a member of the UN-REDD Programme. The first forest carbon inventory of the Terai Arc Landscape was conducted in 2011. A REDD+ pilot project was conducted from 2011 to 2013 in three districts: Dolakha, Gorkha, and Chitwan. The GoN submitted an Emission Reduction Program Idea Note (ER-PIN) to the FCPF in 2014, and an Emissions Reductions Program Document (ER-PD) was prepared in 2015 (Rai, Pant, and Nepal 2018).

The year 2015 was a critical one for Nepal on several counts: a 7.9-magnitude earthquake with Gorkha as the epicenter led to more than 8,000 deaths and infrastructure damaged. The new Constitution of Nepal was promulgated in 2015, resulting in a provincial government with seven provinces and 77 districts replacing the former political structure of five zones and 75 districts. The National REDD+ Steering Committee and National REDD+ Coordination Committee were also formed this year, leading to a restructuring of forest governance, with the 74 District Forest Offices (DFOs) transitioning to 84 Division Forest Offices (DFOs) across seven provinces as well as the formation of provincial and district chapters of civil society groups. Following the endorsement of REDD+ by the Paris Agreement in 2015, Nepal submitted its INDC in 2016, endorsed the Nepal National REDD+ Strategy (2018–22) in 2018, and signed an Emissions Reductions Payment Agreement (ERPA) with the FCPF in 2019. According to this agreement, Nepal anticipates performance-based payments of USD 45 million by 2025 (MoFE 2019a). The Ministry of Forests and Environment (MoFE) was also formed in 2018, replacing the erstwhile Ministry of Forests and Soil Conservation (MoFSC) and Ministry of Population and Environment (MoPE). Climate change was added to the mandate of this new ministry. As per MoFE (2019a), the federal Government of Nepal makes national forest policy, while the responsibility to implement these policies and manage forests lies with the provincial government. It is clear that much of the restructuring in recent years mirrors the priorities of the new REDD+ architecture, given the explicit and clear articulation of Sustainable Forest Management (SFM) as a key priority for MoFE. This is reflected in the National REDD+ Strategy (2018–22) and the Forestry Sector Strategy (2016–25); as well as in the new Forest Act, National Forest Policy, National Agroforestry Policy, and National Climate Change Policy, all ushered in during 2019, another watershed year for new forest legislation in Nepal. SFM was linked to national prosperity in ecological, social, cultural, and economic terms, and was clearly associated with enhancing resource productivity, employment, and revenue while supporting climate mitigation and adaptation (MoFE 2019a).

According to the ER-PD prepared in 2015 and revised in 2018, the Emissions Reduction Program Area (ERPA) constitutes 12 (now 13) contiguous districts of the Terai Arc Landscape (TAL), covering approximately 15 percent of Nepal's total land mass – 2.4 million hectares of Nepal's most productive agricultural lowlands and parts of the Chure Hills – that is home to 25 percent of Nepal's population, with the goal of serving as a model to be scaled up nationwide to address the drivers of deforestation and degradation in Nepal. The agreement was made between the FCPF Carbon Fund and the Ministry of Finance, allowing for results-based payments of up to USD 70 million for 14 $MtCO_2e$ from this program. The ER program anticipates a reduction/removal of 13.2 $MtCO_2e$ by 2024 (26.3 $MtCO_2e$ by 2028) under the Carbon Fund – with provisions for a 23 percent uncertainty buffer (12) and reversal (11) mechanism – budgeted at USD 123 million for a six-year period (USD 184 million for 10 years), into which the GoN also contributes funds of approximately USD 45 million (over 10 years), along with co-financing from various non-governmental programs such as the WWF-TAL Program (USD 13 million) and Forest Investment Program (USD 7.5 million), and from community and collaborative forest user groups (USD 13 million) and household rural energy users (USD 10 million) "through existing cost sharing arrangements" (World Bank 2018, 6). In line with these expectations, the 2019 Forest Act has stipulated new guidelines for community forests: that 50 percent of income be earmarked for poverty reduction, women's empowerment, and enterprise development activities, and at least 25 percent in forest management, and that these and other financial inflows contribute to a Forest Development Fund (MoFE 2019a). USD 51 million is the expected revenue for the GoN over a 10-year period from the sale of 10.2 $MtCO_2e$ to the Carbon Fund at the rate of USD 5 per tCO_2, to be paid in two instalments after years 4 (2022) and 6 (2024) (MoFE 2018, 2019b; World Bank 2018).

Promised benefits for forests and people

Similar to the case of CDM, REDD+ in Nepal does not involve a direct monetary transaction between buyer and seller of carbon credits, hence carbon revenues do not directly trickle down to the forest stewards. Rather, carbon finance is seen as an investment for the sustenance and strengthening of existing forest governance structures, and the benefits are expected to accrue to "communities who, as a result of ER Program activities, are managing their own forests, receiving technical support and extension services from MoFE and are being trained in sustainable forest management techniques to generate higher productivity and revenues from their forests" (World Bank 2018, 5). In tandem with the Nepal REDD+ strategy, the ER-PD claims to seek 'carbon and non-carbon' benefits by implementing seven 'interventions', including the creation of more community forests and collaborative forest user groups (more than 200,937 hectares), 'improvement' of existing traditional and customary management practices in existing forests (336,069 hectares), and expansion of subsidized access to biogas and improved cook-stoves (60,000 units each), among others. It is anticipated that women in particular will benefit the most from the latter; all forest-dependent communities – particularly women, marginalized caste and ethnic groups (Dalit/Adibasi/Janajati) – are expected to benefit from technical capacity-building, and community development from money generated from reduced emissions; and Indigenous and vulnerable groups are expected to be protected from possible risks of disenfranchisement by adherence to the FPIC principle prior to the implementation of ER programs. It is claimed that the ER program was "designed in a highly consultative manner (over 70 consultations in the last two years)" (World Bank 2018, 4).

But the nature of representation and participation of marginalized groups has been strongly refuted by Satyal et al. (2019), who claim that their representation has been symbolic to formalistic at most and their participation has been largely focused on information-sharing rather than being empowered to meaningfully contribute to the decision-making process. They found that while some civil society organizations such as FECOFUN and NEFIN have been relatively more visible in the deliberative process – for instance, being invited to be members of the REDD Working Group – others, such as DANAR and HIMAWANTI have been only peripherally involved (such as in the occasional multi-stakeholder forum) despite their desire to be more engaged. FECOFUN has been already quite active in forest politics and governance, and in the context of climate discourse, their power is receding. NEFIN, on the other hand, has been able to secure a political space hitherto unavailable (Satyal 2018). The disparity in access and engagement between NEFIN and DANAR is particularly telling of the external imperatives driving the impetus for social equity in the country – NEFIN is more connected to the international networks of Indigenous discourse, while DANAR's focus is on caste-based discrimination and oppression that is relevant to the South Asian region. Madhesi and Muslim representation is even more lacking, both in national- and local-level discourse, although their concerns seem to be raised by NEFIN in the deliberative process (Sherpa, Rai and Dawson 2018). At the community level, Dalits are not only invisible but actively discriminated against; at the level of national discourse and policy-making, they feel sidelined in relation to Janajatis; powerful local elites have been largely successful in capturing opportunities generated by the REDD+ deliberative process (Satyal 2018; Satyal et al. 2019), and they have been known to coerce Dalits and Janajatis to participate while begrudging them the provisions for positive discrimination, even as members of these groups, particularly women, are the most affected by constraints on everyday forest practices such as grazing imposed on account of REDD+ (Kane et al. 2018). Thus while the GoN sees REDD+ as a possible new source of investment for Nepal's forestry sector and for its ability to undertake voluntary commitments towards a low-carbon and deforestation-free development pathway, it acknowledges that Nepal's capabilities for compliance social safeguards are still weak. REDD+ advocates in Nepal realize that while, on the one hand, non-carbon benefits for local communities need to be prioritized in order to receive community buy-in, on the other, mechanisms for social safeguards, including FPIC, gender equality and social inclusion (GESI), ILO Convention 169 and UNDRIP, good governance, and equitable benefit-sharing mechanisms need to be strengthened to receive buy-in from funding agencies (Dhungana, Poudel, and Bhandari 2018).

These new developments in forest governance in Nepal in light of the global climate crisis and the particularly ways in which Nepal's forests and communities are implicated signal some paradoxes. On the one hand, similar to a wave of globally created environmental agendas that precipitated radically progressive structural reforms in Nepal's forest governance in the 1980s and 1990s, there is an opportunity for a global climate agenda to nudge Nepal into adopting socially progressive reforms particularly centered on gender and Indigenous rights, as well as to continue to work on better governance, pro-poor, deforestation-free development, and sustainable forest management. On the other hand, there are indications of yet more reversals on the gains made by previous structural reforms and concerns about technocratic domination and centralized decentralization as raised by Pokharel et al. (2008). Meanwhile, the pendulum seems to be shifting back to a state of privatization for Nepal's community forests – having passed through stages of privatization, enclosure, centralized to decentralized governance of common pool resources – given the particular modalities through which REDD+ financing is designed. As other researchers have found from studying the early experiences of forest communities in Nepal with REDD+, there have been tangible financial and social benefits for CFUGs, but they may have occurred "at the expense of autonomous decision-making and customary rights related to forest access" (Poudel et al. 2014, 39). There have been concerns that REDD+ might have been portrayed in too positive a light, downplaying inherent complexities and challenges (Khatri et al. 2016). Likewise, a lack of adequate awareness of the climate change-forest linkages among the local populace even in pilot project sites is seen as an impediment to the equitable practice of REDD+ (Gilani, Yoshida, and Innes 2017; Mandal et al. 2013). Eventually scholars cautiously optimistic about REDD+ argue that for it to work as intended, REDD+ needs to eschew its reductionist approach to "evidence-based payments based on the current market price for carbon" and that a more holistic approach to sustainable livelihoods cognizant of local complexities is warranted (Corbera and Schroeder 2017, 6).

While Nepal has been attractive for REDD+ due to its unique system of CF, therefore, ironically the very things that make a vibrant and sustainable institution with significant local autonomy stand to be eroded due to the need for carbon legibility. On the flip side, if REDD+ is going to be successful in Nepal's community forests, it appears that the only way that may be remotely possible is if it is willing to be flexible on its techno-bureaucratic approach to supporting sustainable management of forests.

REDD+ in Nepal – perspectives of CFUG members

My interest in researching REDD+ in Nepal rose in response to exhortations to oppose REDD+ I had frequently heard in the United States. Having been familiar with the broad contours of CF in Nepal, and learning during the course of my research on CDM that there had been REDD+ preparatory work underway in Nepal, I was curious how responses to REDD+ in Nepal would compare with those among Indigenous groups in the Americas. My research on REDD+ in Nepal spans a period of three monsoons, 2017–19. I met with a wide range of employees of Kathmandu-based organizations in July and August of 2017: the REDD+ Cell/Implementation Center, MoFSC, FECOFUN, NEFIN, Ministry of Science, Technology, and Environment (MoSTE), WWF-Nepal, ICIMOD; visited the three pilot project sites in Dolakha, Gorkha, and Chitwan and met with members of CFUGs, DFOs, and FECOFUN; and attended workshops for CF professionals and members in these places as well as in Pokhara. I returned for follow-up interviews in Gorkha and Chitwan the following monsoon, 2018, and participated in REDD+ workshops in Kathmandu, Nagarkot, and Chitwan. In August 2019, I organized a number of focus groups in the form of several one-day workshops in Chitwan, one in Nawalpur, as well as a three-day forum in Pokhara, supported by grant funding, for 24 CFUG representatives from the 13 districts in the Terai implicated in Nepal's first ERPA with the World Bank. My observations and analysis in this and the next chapter mostly draw on qualitative interviews and participant observation during these interactions, focused on perceptions about REDD+ among forest users. All conversations were in Nepali and I have translated them to preserve meaning, flow, and nuance. CFUG names are provided for Chitwan but not the other districts from where people traveled to Pokhara to participate in the three-day forum. Ethnic markers are used where known, on the basis of either last name or self-identification – they are left out if there was uncertainty during the transcription process. All quotes shared in

Figure 5.1 Premises of the Janapragati CFUG office, Shaktikhor, August 2017

this chapter and the next are from meetings in July and August 2019, unless otherwise noted.

From 'they're going to buy up the oxygen' to 'we have high quality carbon to sell'

> When people in this village [Shaktikhor] first heard about REDD+, they didn't want anything to do with it. They believed "the foreigners wanted to buy up our oxygen". We were absolutely against it. We have since learned that that's not what REDD is about. Nowadays everyone here wants REDD to return.
>
> (Bahun/Chhetri/female, Janapragati CFUG, August 2017)

> Villagers used to say, "Are you selling air?" The year after we received the pilot money it didn't rain. Village elders started to say "every time the foreigners come they take our *jangal*. Now foreigners have taken our *hawapani* [local climate] away". . . . They don't understand, they are under the illusion of their ignorance, say they will take our *hawapani* away. Our village elderly, we tried to explain in simple language: Foreigners pollute, our trees help them and us. This is what we understand [about REDD+].
>
> (Adibasi/Janajati/female, Chelibeti CFUG)

Shaktikhor is a locality in the Kayarkhola Watershed in Chitwan, with 16 community forests covering 2,382 hectares. Kayarkhola Watershed received about USD 21,900 in 2011 and about USD 24,695 in 2012 in carbon payments for sequestering more than 2.5 million tons of carbon above the 2011 baseline. A 2009–13 Forest Carbon Trust Fund (FCTF) project named 'Design of and Setting up a payment system for Nepal's Community Forestry Management under REDD+' sponsored by the Norwegian government was implemented by ICIMOD, FECOFUN, and ANSAB) in the three pilot project areas. Between 2010 and 2013, 105 CFUGs received REDD+ 'seed money' as follows: USD 120,579 for 58 CFUGs in the Charnawati watershed of Dolakha district; USD 69,055 for 16 CFUGs in the Kayarkhola watershed, Chitwan; and USD 73,666 for 31 CFUGs in the Ludikhola watershed, Gorkha. The three watersheds contain 13,970 hectares of forest, of which 10,265 hectares are community forests with 104 CFUGs and a population of 93,000, including Bahun, Chhetri, Dalit, and Adibasi/Janajati groups. Funds were allocated to selected CFUGs on the basis of: amount of carbon sequestered; the proportions of Dalit, Adibasi/Janajati, and poor households; and ratio of men to women (Gurung and Gurung 2018; Satyal et al. 2019; Shrestha and Shrestha 2017). In Chitwan district, Kankali, Janapragati, Jamuna, and Chelibeti CFUGs were among those that received seed money. Amrit Dharapani, Baghdevi, Chaturmukhi, and Jaldevi CFUGs were not, although soil sampling and carbon measurements were conducted in Baghdevi CFUG. Various multi-day workshops on climate change and REDD+ had been conducted in and around the pilot project sites, starting around 2006. Those who had participated in these workshops often shared similar insights about "the factories in the developed countries emitting carbon" and how Nepal's community forests were seen as a haven for storing that carbon. Yet, it was not all that clear to them what that meant. They felt the workshops were too brief and desired more information. Most people found that 'carbon trading' – *carbon vyapar* in Nepali – was a better term to understand than the strange-sounding REDD+, which was often mistaken for the Red Cross.

There was overall a hesitant enthusiasm for REDD+ seen as a financial resource among the majority of individuals I heard from – male/female, Dalit, Adibasi/Janajati, and Bahun/Chhetri – much confusion and curiosity about what exactly it was, and some degree of hesitation and wariness about possible risks. A common refrain was that carbon was a resource going to waste – *khera gairaheko shrot*. REDD+ or no REDD+, everyone valued forest conservation and were determined to continue to steward their forests and to use their resources. If carbon trading brings in resources, this would be welcomed as long as carbon money comes directly to the community, and there were no constraints to traditional harvesting practices, such as cutting grass and grazing goats. Forest users at the grassroots level were seen as most deserving of any monetary or community development benefits. Equitable distribution of benefits within the community was a priority, so were relationships between the community and the state, and the global community. It was strongly felt that since the forest users do the work of conserving forests and animals, they should see the benefits. But there was also a sense of resignation that this is not how things work, and a desire to change that dynamic.

> The [CF] users have conserved the forests. We don't know if these users who have done the work of conserving forests will receive the funds. These are low level users, looking for sources of income. If they are to receive *pratyakshya* [directly] it is good. Forest users don't even know that we are conserving everyone's carbon. Even us elected representatives do not know what is happening. I don't think carbon trading is bad. We have a lot of forest. If the money is received directly at the local level, it is not bad. We conserve carbon. It is good for developed countries to invest in this. Whether or not we trade carbon it is good to have plants. This is an abstract business. Trees store carbon and we get money. If DFO and others can effectively explain to the [CF] user, if they can understand *pratyaksha*, programs can be introduced.
>
> (Bahun/Chhetri male, Jaldevi CFUG)

> It is good for a *khera gairakheko chiz* [something that is wasting away] to bring revenue. The government has already decided 1–2 years ago. There were discussions and debate, to sell or not to sell. We were mostly in agreement . . . government can proceed. But it cannot proceed only with agreement at the topmost levels who only see money. Affected groups must be consulted. The involvement of those who tend the forest must be secured.
>
> (Dalit female, forum participant from Rautahat)

> As far as I understand, *Jal, Jangal, Zameen* are things we cannot live without. Forests help against climate change. They are valuable. Our ancestors had used up forests. Their value was realized and so we have [CF]. *Hariyo Ban Nepalko Dhan* [Forests are Nepal's wealth]. REDD+ is a business. The more greenery in forests, the more carbon there is. The carbon that is going wasted brings financial resources. But conserving forests is also our *Dharma*. It is mentioned in the *Ved* and *Puran*. It brings money, helps conserve, it's good for our health. There are only benefits. If the government arranges for this to happen, it would be good.
>
> (Adibasi/Janajati male, Janapragati CFUG)

Many were not confident that benefits would be equally distributed, whether in the community or within the community of CFUGs. A common refrain was that those with *pahunch* (connections) usually nab the resources and opportunities, including participating in workshops. CFUG members in a neighboring district of Chitwan, Nawalpur – outside of the pilot project areas – who had heard about REDD+ on the radio and through word of mouth, asked why they had not received a pilot project even if their community forests were thriving. CFUG members in Chitwan who were outside the pilot project areas shared these sentiments as well, wondering if stronger FECOFUN connections had played a part in Shaktikhor being chosen. A FECOFUN representative was not sure why they were chosen but agreed that FECOFUN was a key stakeholder due to the structure of their organization, a nationwide federation with branches at each scale of governance (Interview, FECOFUN member, Bahun/ Chhetri female, Shaktikhor). There seemed to emerge a sense of competition and animosity among CFUGs as to which forests had more carbon and were entitled to benefits. A commonly expressed concern in interviews with the more knowledgeable CFUG members was also that resources when they trickle down to the local communities not be diverted to middle persons, whether in CFUG/FECOFUN leadership, or in centralized bureaucratic channels. Even if communities did receive training, they resented their formalistic/obligatory nature. DFO officers who get training in REDD+ get shifted to other districts and their training is wasted. There are similarly frequent changes – every three years – in CFUG leadership and the knowledge is typically not passed on. There are a lot of unknowns: what is carbon, how much is there in a tree, how to measure, who owns the carbon in the soil, who owns the carbon in the forest, and what discussions are occurring in the international context. Several CFUG chairs said that even as chair they didn't understand REDD+ fully, and they desired more training and local capacity-building.

> All the legwork for carbon measurements were done by the locals: digging soil in depths of 10, 20, 30 cm, measuring tree diameters. . . . People get 300 rupees. And the salary of Kathmandu-based experts. We have to talk about this. . . . Experts from Kathmandu and abroad came and used some fancy gadgets and they earned a lot of money. They should have given it to us. . . . If instead of hiring them we invested in enhancing our own capacity, that money would go far. Everyone could be taught how to map carbon. We are the ones who do the hard work. The expert only comes and uses the GPS. How hard is that to do?
>
> (Bahun/Chhetri female, Janapragati CFUG)

> There are workshops but most people don't know about them because the chairs only invite those they know, or only the chair and secretary go. Leadership of CFUG has had some opportunities to attend workshops and understand a little about REDD+. The ordinary forest user who collects ferns and other products, the one who is most at risk from wildlife, he doesn't get to participate, and he doesn't have time anyway. That's why it is not well understood by the ordinary forest users, the ones who do the actual work of conservation, management and disaster alleviation.
>
> (Bahun/Chhetri female, Chaturmukhi CFUG)

> The workshops are not effective because they only last for a day or several hours and the learning is not deep; they are often conducted in a rush toward the end of the year's closing of the budget. Concepts are never

explained or understood well, and are forgotten easily. 2–4 day course is needed for thorough teaching and learning.

(Adibasi/Janajati, Dhuseri CFUG, Nawalpur)

In any case, the injection of funds in 2011 and 2012 was welcomed by those who received them in the pilot project areas and had created intense hope among others nearby or those who had heard about them. Already, CFUG funds were being channeled towards a wide range of community development projects – including for education, recreation, livestock rearing – and there was excitement about new and additional support from REDD+ to sponsor more of these projects. At times, frustration had set in for those who had waited many years for a new influx of funds, and they were not sure why it had not come. Members of Janapragati CFUG, who were enthusiastic in 2017, seemed somewhat jaded by 2019, and they had no knowledge of the ERPA, even if their district Chitwan was one of the 13 that were included. One particular gaffe made during the pilot project was unintentional – a well-meaning gesture with unintended consequences – when targeted CFUGs were led to believe that funds were being disbursed as a 'financial reward' for increasing carbon stock, without adequately explaining that future payments would be more transactional – thereby creating unrealistic expectations and false hope about the future (July 2017 interview with ICIMOD official, Bahun/Chhetri male, Kathmandu). Several people in Chitwan had learned of this and said that earlier expectations of making windfall profits from the sale of carbon were diminished. Even if they did not understand the details, they were vaguely aware of the injustice of climate change and of the risks involved as well as the notion of unequal exchange in the carbon transaction, yet most of them wanted to attract REDD+ investors and to seek ways to maximize financial benefits while minimizing risk and protecting their rights. While some people in focus group settings suggested critical insights or raised questions about the risks, such sentiments were typically overpowered by the group's consensus. One male participant at the forum, chair of a local FECOFUN chapter, put it this way:

We can't just be suspicious about everything. . . . It's our right too. . . . We should seek to learn from the pilot project and move ahead with carbon trade with agreement from CF and with good governance and FPIC. . . . We should not let go of carbon trading, we should embrace it. This is like eating the honey that bees make. It is like *baaf line* [inhaling steam]. We should strategize financial benefits without compromising rights, without becoming slaves. They will try to enslave us but we won't relent. . . . The US has bullied everyone. . . . Its hegemony is everywhere. We have to keep raising our voice. We have to seek alliances with India and China. They have a powerful voice and they can speak for us and help strengthen our position. . . . Since 1816 Sugauli treaty Nepal has always lost in foreign relations with India. But in international treaties Nepal may have an advantage. Carbon trading is international. Nepal should ally with countries such as Brazil and forge ahead, involve political parties and get them to understand the issues.

(Bahun/Chhetri male, FECOFUN, Pokhara)

This level of enthusiasm is particularly interesting because the understanding of REDD+ and climate change itself is rudimentary and convoluted. One of the participants (from Janapragati CFUG in Shaktikhor, who had relatively more experience with REDD+ programming) described climate change as a problem of ozone thinning. To what extent community buy-in is a result of an incomplete or inaccurate understanding of the link between climate change and forests, is unclear. A similar instance revealing (high interest coupled with) just how misunderstood the problem of climate change and its connection to forests is, occurred during a conversation with a former CFUG leader and elected local official in Dolakha, where she insisted that her community forest had "high quality carbon to sell" (Interview, July 2017). As in the other pilot sites, her community had benefited from the significant influx of funds and eagerly anticipated more. But Dolakha and Gorkha districts had not been included in the ERPA. I did not return to Dolakha after 2017, but when I spoke with CFUG committee members in Gorkha in 2018, I learned that they had not heard the news yet, and now wondered if they did anything wrong and therefore were not selected. Ironically, the rationale for choosing the Terai for the first REDD+ program was the higher emission reduction potential there, since the hill and mountain regions had a better record of forest management, and less room for growth (Interview, Bahun/Chhetri male, ICIMOD, July 2018).

Free, prior and informed consent: state-society disparities

Heard that European countries were going to bring millions. Developed countries have a lot of industrial pollution and are seeking out forested countries. Oxygen is good, carbon dioxide is bad, but you can sell carbon dioxide. Why would anyone want to buy something that is bad? We have heard on the radio that even the international community is using our oxygen.

(Male, Jaldevi CFUG)

We might be blinded by visions of cash flows. But what is behind it? What if we are not able to use the resources we conserve? If we can then it is okay. But not if we lose jurisdiction over our forests. If after accepting money they place constraints on our resource use, we will suffer. It's about us helping them deal with their pollution and getting compensated for it. The forests do it on their own. But we don't want them to take away our rights over our forest. . . . We need to understand what we are getting into. What, who, where, how? Who buys it, and where? Who will measure carbon? We haven't understood it well. We have to understand all these issues well or tomorrow we will regret.

(Adibasi/Janajati, female, forum participant from Rupandehi)

Figure 5.2 Laxmi Mahila CFUG, Gorkha, August 2017

> Humans are resourceful beings. They have invented many things, achieved high levels of development. They can destroy the world in no time with a bomb. Now they are talking about carbon farming. Foreigners should set aside a budget for this, that's good, if done fairly. But how will the carbon funds be distributed? In a bundle or disaggregated? What will happen if the carbon stock goes down for some reason, drought, fire, storms, etc. Will there be consequences? We have to think of all this.
>
> (Adibasi/Janajati male, forum participant from Dang)

Such incomplete understandings and mischaracterizations of forest–climate linkages reveal not cognitive deficits on part of forest communities but rather that information about REDD+ had not been clearly or adequately communicated by state and civil society bodies. As such, they raise serious questions about whether the principle of Free, Prior, and Informed Consent (FPIC) so ostensibly integral to REDD+ has been adequately followed before the ERPA was finalized. The focus groups in Chitwan as well as the three-day forum indicated that knowledge and understanding of REDD+ was uneven. A few people said they had never heard of REDD+. Most had heard about it, on the radio; some had read about it in newspapers or through FECOFUN over the past decade. Some of them had participated in workshops about REDD+ and/or climate change. Some CFUGs in Chitwan, such as Janapragati and Chelibeti in Shaktikhor that had greater familiarity, had benefited from workshops on climate change, gender sensitization, carbon measurement training, and received financial resources. Even among those with exposure, misunderstandings were clearly rife about the complex linkages between climate change, forests, and the role of REDD+. Importantly, REDD+ was not seen as a transaction, but rather as a reward; thus, the notion of negotiating a price for carbon being sequestered was largely absent, and any understandings of the forest-climate linkages were understood in the local context, with the global nature and drivers of the problem largely missing. Further, information dissemination about the process of REDD+ implementation seemed to be lacking as well, since neither Gorkha nor Chitwan-based CFUG members had received news about the ERPA, which community forests were or were not selected, and why.

Critical questions that were emerging in policy documents and discourse – such as who owns the rights to carbon sequestered in trees and in the soil, how liabilities such as forest fires would be handled, and what are the conditionalities of the ERPA – were not part of the discourse at the community level. Community engagement seemed to be mostly focused on how funds once they come in would be disseminated equitably within the community and invested in sustainable forest management. As such, perceptions of REDD+ among forest communities were mostly framed as an economic development opportunity, and a professional development or capacity-building opportunity. *Who could be against these?* At the end of the three-day forum, participants were invited to work in break-out groups to discuss and articulate their key priorities. The following is a distilled version of what they came up with, explained in more detail in a white paper shared with the MoFE and the World Bank among others (Joshi 2019).

1   Right to accessible, relevant, and accurate information and communications about REDD+; informed consent from all affected CFUG members, particularly the most marginalized and vulnerable, before irreversible agreements are made at the international level.
2   Meaningful and equitable participation and representation in terms of gender, income, caste, and ethnicity at all stages of decision-making, implementation, and benefit-sharing, facilitated by empowerment of CFUGs for carbon measurement and ownership of forest carbon.
3   Fair and equitable rules – policy and laws – for carbon trading with meaningful participation at all levels, particularly from local representatives, along with accountability for central level bureaucracies; CFUGs retain traditional resource rights, as well as carbon rights.
4   Equitable benefit-sharing within CFUGs according to gender, ethnicity, and caste, with effective and efficient distributive mechanisms; forest users have a say in determining minimum price of carbon, and not be reduced to serving as hired soldiers for the protection of their forests.
5   Risk management provisions for local communities if unanticipated and intractable problems arise, e.g. insurance policies against forest fires.

In response to this white paper, a MoFE official agreed that the Cancun Safeguards principles emphasized by the participants at the forum – particularly FPIC, meaningful participation of local actors, and gender inclusion – were indeed important and should not be overlooked but indicated that there were financial and other constraints in involving all stakeholders in state law and policy formation (Email correspondence, September 2019). Indeed, from the vantage point of the state, given the scale of community forestry in Nepal, involving thousands of CFUGs and their members in everything from information dissemination to involvement in decision-making can be a daunting task. To further complicate the charge, climate change is a complex phenomenon to understand and to explain. When you add linkages with other systems such as forests, as well as the complexities and technicalities of a performance-based compensation scheme, particularly in a still-evolving landscape of international climate governance, the level of complexity ratchets up significantly. There is also a tendency among Western and Western-educated specialists to assume a deficit-based approach towards the capacity of rural resource users to understand technical and complex subjects and therefore to have a paternalistic orientation where decisions are made unilaterally. But my research revealed a strong desire for knowledge of how carbon trading operates in the local, national, and international contexts; technological support; and involvement in carbon measurements so that resources could remain in their communities rather than be diverted to outside experts.

Community forest users wanted to be active participants, not passive recipients in Nepal's REDD+ programming. Without a proper understanding among those who would be most affected by REDD+ of how it operates, how it strives to mitigate climate change, and how local forests and forest users are implicated, I argue that the FPIC safeguard has not been rigorously followed throughout the CFUGs in Terai before the ERPA was signed. The existing array of organizations and institutions seem to have neglected making this a top priority. It is not that training and workshops have not been conducted; they have, but it appears that the target audience for these educational efforts have tended to be the topmost officials at most bureaucratic organizations, with no obligations for passing down information and knowledge to their constituents. Part of the problem is that capacity-building workshops are often conducted in central locations or city centers such that rural forest users, even when they receive opportunities to participate, must juggle family and work obligations to travel to the workshop or training venue. Such travel requirements especially create constraints for women to participate. Due to these many reasons much misinformation and information gaps persist between the state and forest-based communities. A coordinated drive to involve local forest stakeholders (users, CFUG members, FECOFUN officials) in all REDD+ districts, through training and workshops about the complex connections between climate change and forests, how REDD+ operates, what the rights of local forest users are, and how they can best advocate for their rights, seems to be an urgent priority. One state bureaucrat's perspective was quite different:

> It's better not to focus on carbon and the political economy question of who wins and loses, the power dynamics. It is not productive. They are already engaged in community management. If you can convince them to do more that will be the most appropriate strategy. As soon as you talk about carbon, the question becomes political. Political questions don't have technical solutions. If you compare timber and carbon, of course timber wins. You cannot justify carbon trade. What we want is to manage the forests with silviculture for timber provisions as well as yield carbon co-benefits. We cannot focus only on carbon.
>
> (Former Ministry/REDD+ official, Bahun/Chhetri male, Interview, Kathmandu, August 2019)

Winners and losers within community forestry

Nepal's engagement with REDD+ reveals gaps not only between the state and CFUGs but also within and across CFUGs. Disparities exist by geography and carbon mitigation potential – hence only one of the three pilot project sites is part of the 2019–24 emission reduction program – but also across CFUGs. The pilot project revealed that CFUGs would compete for funds based on carbon stock enhancement and demographic criteria. Within each CFUG that does receive payments, there are pre-existing power differentials along gender, caste, ethnicity, and wealth lines – that co-exist at multiple scales – and intersect and evolve in new ways due to the REDD+ agenda.

Gender

One of the most visible disparities from my point of view as a researcher was that of gender. This is a multi-faceted and multi-scalar issue with reverberations in resource access, participation constraints, and cultural domination. My Kathmandu-based interviews with individuals knowledgeable on REDD+ were again – as with CDM research – overwhelmingly with male officials, including in organizations such as NEFIN, FECOFUN, ICIMOD, WWF, and relevant ministries. With the exception of NEFIN, they were also predominantly Bahun/Chhetri males. The DFOs I interviewed in Gorkha and Chitwan were also Bahun/Chhetri males; the only women within the Chitwan DFO setting were five of the six Local Resource Persons (LRPs) who were hired on contract to conduct fieldwork in carbon assessments, a disparity that a male officer acknowledged with regret was a structural flaw; the LRPs were desirous of more work having received training, but recognized how contingent work opportunities were, and they had been paid a mere Rs. 500 per day of considerable legwork; one of them also shared that during fieldwork, their insights were often dismissed by the 'Sirs' who would tell them what to do, even if they (LRPs) were the ones who actually understood the 'ground realities' better (Interviews, female LRPs, August 2018, Chitwan). In 2018–9 a female forestry scholar with expertise in gender was appointed as the REDD+ Implementation Center chief in Kathmandu. This was clearly an exception to the rule but appears compatible with the state's desires to incorporate Gender Equality and Social Inclusion (GESI) more centrally in national planning and strategy development. Similar institutional reforms were introduced in CFUG administrative structures as well, starting in the Forest Sector Master Plan of 1986–7, now requiring women to hold at least 50 percent of executive committee membership, including in its top two positions of chair and secretary. In practice, women's presence in decision-making positions in CFUG administration has gone up to 25 percent since then (Gurung and Gurung 2018).

While creating space for female bodies is a welcome move, it does not easily translate to equally shared power, in a social climate of entrenched patriarchal mores. In fact, the quota directives appear to have a tokenizing effect, where women are thrust forward when outsiders visit – similar to the coerced 'participation' documented by Kane et al. (2018) for Dalit/Adibasi/Janajati members – but mixed-group interactions clearly reveal a culture of male domination, in terms of taking up space, projecting voice, opinions, and knowledge, and mansplaining tendencies. Women typically do not speak openly in mixed gender settings – although sometimes women CFUG executive committee members would be put on the spot and sometimes were told to say (in not very subtle ways) they were 'empowered'. Interviews with female CFUG office-holdersconsistently revealed that knowledge and exposure to REDD+ was typically less relative to male office-holders – in part an outcome of differential constraints and opportunities for attending workshops, particularly in distant cities (Joshi 2019). Yet, CF has indeed created opportunities for women's leadership to thrive. An excellent example of this is embodied by the first and only all-women CFUG in Chitwan, aptly named Chelibeti (translates to womenfolk and daughters), whose leaders describe the experiences of marginalization, silencing, and de-prioritization of women and their issues in CF that led them to form their CFUG when an opportunity arose to steward a 68.5-hectare stretch of forested land in 2009 when the erstwhile 16-year-old DeviDhunga CF was deemed by the Chitwan DFO to be over the 199-hectare limit. When I met members of this CFUG in Bharatpur, they had just visited the DFO to express in writing their discontent over the new 2019 Forest Bill's tax provisions.

> The work is not easy but we needed this. . . . No matter how accomplished we are, when we are in a mixed group with men we tend to hand over the decision-making power to men and simply agree to follow their lead. So I never make decisions. I give it to them and just applaud their decisions. So what happens is we have an internal dialogue – can't *we* make the decisions? Can't *we* manage the forest? If the males can make policies why can't

we? We are bothered by these thoughts. . . . Once we created Chelibeti, we had to do everything ourselves, solve problems, make decisions, liaise with DFO, run the school. Us women didn't know what "cufeet" [cubic feet] was, even after creating the executive committee. We didn't want to seek help from the males in case they taunted us for our desire to keep the men out of it. We had to eventually learn about selling timber outside the community, learn the jargon, how to measure width, circumference, height. There were many sleepless and anxious nights. We didn't have to seek help from men in the community. . . . The DFO challenged us to be bolder. Women already are shy to speak in public but we had to speak into a microphone at the first general assembly. We used to tremble at the thought, but we did it. . . . We learned eventually that no matter if it is a big or small CF, the policies are the same. Anyone can do this work. We now have accumulated experience of doing it. We feel proud. Until we experience it we used to think we couldn't do it, that we didn't have decision-making ability, we wouldn't be able to face challenges, but that turned out not to be true. We now have experience.

(Adibasi/Janajati female, Chelibeti CFUG)

Chelibeti was one of the CFUGs that received REDD+ pilot monies (Rs. 82,000, Rs. 68,000, and Rs. 90,000) as well as training in carbon measurements, which they were eager to expand and put into practice for additional income. But most other CFUGs appeared male-dominated despite the 50 percent rule. Gurung and Gurung (2018) similarly argued that Nepal's efforts to institution-alize a GESI agenda has not been effective enough despite strides in greater representation in numbers, since existing cultural and social norms continue to keep decision-making positions out of reach: they found that when women tried to run for chair of the CFUG executive committee, they faced resistance from established male rivals and the community at large. A similar account was shared with me in the context of a visit and interviews at Chaturmukhi CFUG in Gaduwa. This explains why despite the 50 percent rule, it was typically the male who was chair and the female who was secretary. The level of resistance to women's leadership enhancement and power-sharing was particularly acute when opportunities for resource or financial in-flows presented themselves. In addition to resistance from society at large, women also experience hurdles within the household, with limited time to dedicate to work outside the home, and moreover, facing discouragement or lack of trust from husbands regarding work outside the household, including attending workshops in Kathmandu or Pokhara. The wide array of women I interviewed in CFUG, FECOFUN, DFO, MoFE settings were keenly aware of women's subordinate roles but exercised their agency in various ways to maximize opportunities within the structural constraints of patriarchal culture – whether by seeking opportunities for extra income, travel, capacity-building workshops for themselves or by strengthen-ing economic opportunities for their organization. Shah (2018) has described how in seemingly apolitical developmentalist interventions, subaltern women

Figure 5.3 Member (Tharu) of Amrit Dharapani CFUG inaugurating a CF-sponsored professional development event, Chitwan, August 2018

can creatively appropriate the external agency's agenda to open up spaces to advance their autonomous interests. The women I met clearly demonstrated their desire for more opportunities for professional and economic growth and seemed to welcome intervention from outside organizations to help raise their relative positions in their local context. Gurung and Gurung (2018) similarly suggest that a multi-pronged, multi-level approach to reinforce gender-sensitive programming appears necessary if women's participation and leadership is to be enhanced at the level of community development and CFUG governance. It appears that national and international interventions can help subaltern women jump scale and overcome the discriminatory and oppressive conditions they often find themselves in.

Caste/ethnicity/Indigeneity

> We were lucky to be one of the 11 [CF]s to receive money at first. The funds came specifically for the poor, Adibasi/Janajati and Dalit. We had to give them 80 percent, for investment in biogas, pig husbandry, fish farming, goat rearing, poverty alleviation. It is not fair. They got things too easily and took it for granted. . . . Some of them are industrious, others aren't. They are not very smart. Their awareness

> levels are low, and education too. . . . They weren't even able to raise the animals and sold them instead, said they had to be fed too much. The money was wasted on them. It should have been given to the whole community. We were preparing for a revolving fund, but the second instalment never came. They did not explain why.
>
> (Bahun/Chhetri female, Janajagriti CFUG, Chitwan, August 2018)

During a focus group with a group of women CFUG members, this complaint was shared by a woman and others agreed with it. It is a paradoxical situation. The higher proportion of poor and marginalized caste (Dalit) and Indigenous Nationalities (Adibasi/Janajati) is one of the reasons that certain watersheds were chosen for the REDD+ pilot project, in addition to carbon mitigation potential, in keeping with the Cancun Safeguards. Specifically, Chitwan was identified as a district with the largest population in Nepal (28,989 in 2011) of a highly marginalized Indigenous group, the Chepang (World Bank 2018). Accordingly, benefits were also expected to trickle down to the same groups. Much of the preparatory work during the REDD+ preparedness work at the local level was focused on building CFUG capacity-building for equitable and transparent resource distribution systems. In order to prevent elite capture and make sure that funds contributed to lifting up livelihoods and reducing poverty for the most marginalized groups, CFUG executive committees were instructed to allocate REDD+ seed money among households according to the following representational criteria: poorest households, 20 percent; women-headed households, 15 percent; Dalit households, 15 percent; and Adibasi/Janajati households, 10 percent, with the remaining 40 percent to be allocated for forest and environmental programs (Shrestha and Shrestha 2017). This quota system – similar to affirmative action in the United States – was seen as unfair by the women at the focus group. This suggests a limitation of a social safeguard strategy that is overwhelmingly focused on a logic of distributive equity – imposed from the top down – that does not attend to participatory or representational equity, and relatedly, the recognition of continued structural marginalization of Janajati and Dalit groups.

FECOFUN, LRPs, and CFUG members based in Shaktikhor shared their understanding that the predominance of the Chepang and Dalits in the area was a key reason for the selection of Kayarkhola Watershed for the pilot project. Yet these groups were absent in CFUG leadership. As the earlier quote reflects, they are an othered group whose ways of life are distinct from other forest users in the area. Sherpa (2017) notes that the seed money was given to the nomadic Chepang for the specific purpose of investing in biogas and animal farming, and as such was a culturally inappropriate imposition that failed to benefit them. None of the CFUG and FECOFUN members I spoke to in Shaktikhor were Chepang. As with Chelibeti, the Chepang have started their own CFUG groups in the Chepang village just north of Shaktikhor, but none of the Chepang CFUG leadership in that area had heard about REDD+.

The identification of Adibasi/Janajati/Dalit groups as intended REDD+ benefiaries has clearly not been effective. This is not to say that the redistributive measures are not important, but rather that a sole focus on it is likely to create social discord, animosity, othering, and strain social relations.

> Whoever's forest has more carbon will get more money. This is very good . . . It should be proportional as per population and carbon. . . . It should be divided by population. It is not fair to choose CFUGs with women, Janajati/Dalit. Big forested areas with fewer populations would benefit as a result.
>
> (Bahun/Chhetri female, Janapragati CFUG)

This was a representative of a CFUG that had received seed money during the pilot project. She felt very strongly about the importance of REDD+ benefits to be scaled to the population of a community on the basis of carbon sequestered per capita, rather than the proportion of marginalized groups in the community. She warned that if benefits were scaled to the population of women and Janajati/Dalit, conflict and strife would inevitably result between neighboring CFs with different demographic compositions. Whereas in the past people would pitch in to help each other during forest fires, competing CFUGs may become disinclined to remain neighborly. While recognizing groups for receiving equitable shares is noteworthy, the issue of underrepresentation of Adibasi/Janajati and Dalit in positions of authority, leadership, and decision-making, and the disproportionate predominance of Bahun/Chhettri males, remains pervasive. This was evident to me through my interactions with a wide range of CFUGs and as well as other REDD+ related organizations, and such patterns have been documented by others as well (Kane et al. 2018; Satyal et al. 2019). Such underrepresentation, in conjunction with external directives and pressures to enact social safeguards, often leads to tokenization. Requests made to CFUG/FECOFUN leadership for introductions to forest users for interviews in several occasions resulted in meetings with Adibasi/ Janajati and Dalit individuals – sometimes but not always also women – who were often not as well-versed in REDD+ related developments in their region.

> I don't understand. This REDD brings financial resources from developed countries. Is it true that Janajatis should be prioritized? What is this? What is mentioned in the agreement? . . . It seems they only want to show our faces, take our photos.
>
> (Adibasti/Janajati male, Dhuseri CFUG, Nawalpur)

> Instead of coming straight to the community, the government works through INGOs, FECOFUN. The benefits never reach the ground level. They do some training seminars workshop just for show. It is just for show, to get signatures. . . . Eventually by the time they come to the ground level, there is very little. That is why these projects are

not effective . . . but they document it so well. The World Bank and foreigners see this documentation. They have language hurdles. Even you don't speak that fluently. A translater mediates our interactions in a way that the foreigner is impressed. That's why the system is broken. We need proper in-depth workshops at the ground level. What is carbon, what does it do? What is carbon, how does it effect global warming? To what extent have we assisted developed countries to absorb the pollution created by them? Otherwise programs are not effective, it is just *guff* [small talk]. Material should be taught in Nepali, too, then everyone will understand. REDD is in English. And then there is the plus, why was it added? This will only be effective if it comes straight to the community level.

(Dalit male, Dhuseri CFUG, Nawalpur)

Improvements in the realm of social safeguards in terms of caste/Indigeneity/ethnicity may benefit from the reserved quotas as for women in political representation, in addition to the distributive equity provisions in benefit-sharing. But this tends to be a contested and debated issue, as was revealed during one of the sessions of the three-day forum, when a lively debate ensued as to whether Dalit status was a reliable predictor of marginality in the contemporary context of community forestry in Nepal, and therefore whether progressive forest policies should privilege caste or income/poverty in seeking to make benefit-sharing equitable. There was no consensus at the end of this discussion, which mirrored 'race or class' debates that have materialized in EJ scholarship based in the United States (Mohai, Pellow, and Roberts 2009). It was also noteworthy that issues pertaining to Indigeneity did not organically emerge in this discussion even as Adibasi/Janajati participants were present at the forum. The latter were mostly silent on REDD+ discussions and spoke about ongoing tensions with the national parks, as they tended to live in the buffer zones adjoining the national parks. National parks are not directly connected to the emission reduction agreement, although their management is included in the non-carbon benefits. The insights of these participants on REDD+ drew on their experience of disenfranchisement when national parks were created, and that continue to the present day.

> Buffer zone signatures were deceptive. If they had known about the constraints, they might not have signed. The Dalits and Janajatis that were supposed to get compensation still haven't. 2–5 people die per year but haven't gotten a penny. What if carbon is like that too? Will it reach the community? Will it be like having an orange orchard but never getting to enjoy the oranges? Listening to this dialogue, I feel as if you are all on one side and I am on the other. I am *peedit* [oppressed]. Today I received a phone call. It hasn't been a month since we planted wheat but the elephants have arrived. We have to borrow an elephant siren to chase them away. That is our life. If we had known it would be like this we might not have signed

the papers. I hope carbon trade benefits the community and the Janajati and Dalit who do the conservation work. Those in leadership positions who bring the papers to sign should explain everything to the community members.

(Adibasi/Janajati female, forum participant from Parsa)

The forest-based Tharus are being disillusioned. Our rights have systematically eroded. We have been squeezed out and pushed to the margins. Collaborative Forest Management is further terrorizing our communities. Should we only suffer losses, never see any profits? We are not getting information, our rights are being suppressed.

(Adibasi/Janajati male, forum participant from Bardiya)

Indigenous people in Nepal have long been disenfranchised by new settlers, but Indigeneity has not been as salient an issue as caste in public discourse. It is an emergent discourse that has enjoyed greater salience due in part to its prioritization in international programs such as REDD+ and Nepal's ratification of treaties such as UNDRIP, and the ILO Convention 169. The Cancun Safeguards and its emphasis on FPIC as a right of Indigenous peoples are grounded in these international agreements and form a crucial element of the global REDD+ framework, and consequently Nepal's REDD+ strategy as well. The institutionalization of these provisions for Indigenous rights seems to have bolstered Indigenous politics in Nepal in a way that even the Constitution has not, and as such, may enable Indigenous activists in Nepal to 'jump scale'. Indigenous leaders in Nepal have been engaged in the REDD+ consultative processes and see the state's efforts to implement these international commitments towards upholding Indigenous rights in Nepal as still inadequate but are supportive of REDD+ and see it as "an attractive prospect for advancing Indigenous interests" (Sherpa, Rai, and Dawson 2018, 72). While the Kathmandu-based leadership of organizations such as NEFIN is well-integrated into national climate discourse, they realize that "effective engagement and meaningful participation is still a challenging task for community leaders because awareness about REDD+ at the community level is still low" (ibid, 76). My conversations with NEFIN members in Kathmandu and Chitwan confirm this assertion about the gap in knowledge about REDD+ between the center and the periphery even within NEFIN. Meanwhile, as Satyal et al. (2019) indicated, there is a marginalization and invisibilization of Dalit issues relative to Adibasi/Janajati issues in Nepal's national REDD+ consultative processes, although in discourse at the CFUG level, Dalit issues predominated. While as with any external intervention, REDD+ enables marginalized groups greater leverage, a lack of sensitivity on part of REDD+ advocates to local dynamics is likely to worsen social relations.

Since the legitimacy of Nepal's REDD+ program rests on its commitment to Indigenous rights, all advocates of REDD+ in Nepal should be held accountable to the following promises, with the understanding that Indigeneity in Nepal intersects with caste, gender, and wealth disparities.

> The ER program implementation will fully respect Indigenous peoples' rights over the land they have been managing traditionally within the scope of governments' legal provision.
>
> (World Bank 2018, 98)

> Free, prior, and informed consent (FPIC) is essential for any REDD+ program design and implementation in the context of Indigenous peoples. FPIC helps to ensure that potential impacts on Indigenous peoples will be considered in decision-making processes for those programs or projects affecting them . . .
>   Free: Independent process of decision making;
>   Prior: Right for Indigenous peoples to undertake their own decision-making process regarding any project that concerns them before its implementation;
>   Informed: Right to be provided and to have sufficient information on matters for decision making; and
>   Consent: Collective and independent decision of impacted communities after undergoing their own process of decision making.
>
> (ibid, 132)

North–South power differential as a double-edged sword to disrupt gender and ethnic inequality in Nepal's community forests

Nepal's forests and forest-based communities seem to be going through a second wave of West-driven structural changes, in the current guise of REDD+. Perhaps in response to a concerted movement opposing carbon trading, REDD+ policy appears to be quite explicitly focused on enhancing environmental justice in the local communities, as the explicit emphasis on social safeguards demonstrates. It is important to note that while distributive equity in local communities is being strongly emphasized by REDD+ proponents, participatory equity has not been as strongly prioritized. There has been an effort to recognize groups that are traditionally disenfranchised, and efforts have been made to lift them up through REDD+ incentives. Due to the imbalance between distributive and participatory justice priorities, the gains for both gender- and ethnicity-based equity are more limited. Looking beyond the community level, significant gaps in power still exist. This includes the power differential between Kathmandu-based agencies and rural CFUGs, on the one hand, and, on the other hand, between Global North investors and a Third World country with development priorities and carbon mitigation potential.

At the level of North–South power dynamics, REDD+ does not seem to be compatible with a climate debt or climate justice imperative, where

Global North countries are meant to take responsibility for mitigating climate change. Rather through an international system of trade, the unequal exchange between core and periphery is perpetuated. To complicate matters, it is not clear that REDD+ in its current form in Nepal constitutes carbon trading as we know it. Different even from CDM, REDD+ appears to be closer to bilateral aid, with the caveat that financial flows will be conditional upon verified results – although climate mitigation is the ultimate goal, non-carbon co-benefits are added for good measure – making the relationship between the would-be buyer and seller more of a client-patron relationship. For detractors of carbon trade, is this arrangement more desirable? I am not sure. But in my mind, the power of the beneficiary is even less than that of the seller, because the power to negotiate has now disappeared from the equation. The Government of Nepal rationalizes that non-carbon co-benefits, both in terms of environmental outcomes and social safeguards outweigh the 'price of carbon', negotiated at a mere USD 5 per ton of carbon equivalent. But the terms are set by the 'buyer'. This seems like an unfair bargain, but it would not be the first time such a deal has been made.

The Government of Nepal has a long-standing agenda to support community forestry in Nepal (MoPE 2016). Whether the REDD+ agenda serves to strengthen or weaken the existing community forestry infrastructure is an open question that further research might reveal. Some researchers have indicated that community forestry and REDD+ are compatible agendas and can create co-benefits and equitable distribution of carbon revenues if proper safeguards are instituted, even as gender and social relations take time to evolve (Aryal 2015; Satyal et al. 2019; Sharma et al. 2017), but fear of the loss of autonomy is palpable among forest users. While they seem to be open to the idea of selling the carbon sequestration potential of their forests – based on incomplete and potentially misleading information – they are resolute about drawing the line where loss of access and autonomy is concerned. On the positive side, just as the first wave of forest-oriented development aid helped usher in a community-oriented political economy of forest governance in Nepal, it appears that advocates for gender equality as well as Dalit and Adibasi/Janajati rights are looking to REDD+ as a catalyst for strengthening the provisions Nepal has committed to in paper but not yet operationalized for making communities less hierarchical and more equitable. It appears that Global North hegemony over the Global South can have a beneficial effect for subaltern groups in the Global South by arm-twisting willing and eager states in the Global South into attending to the needs of their subaltern populations. Similarly, state authorities could help the most marginalized in the local context withstand local and regional bullies – a common occurence in CFUG contexts (Gurung and Gurung 2018) – with radical reforms that can restructure power for genuine distributive and participatory justice outcomes. But this is only as true as the intentions behind the social safeguard measures are genuine and not a façade to appease the detractors of carbon trade. While this risk is always present, forest-based communities have demonstrated that they are not willing to passively let the state or foreigners run roughshod over their forests and their rights. This has been a centuries-long power struggle and it continues.

References

Agrawal, Arun. 2005. *Environmentality: Technologies of Government and the Making of Subjects.* Durham: Duke University Press.

Agarwal, Arun, and Sunita Narain. 1990. *Global Warming in an Unequal World.* New Delhi: Centre for Science and Environment.

Aryal, Pabitra. 2015. "From Green to REDD-Ready to Trade: Discussion on REDD in Nepal's Community Forestry." *International Journal on Environment* 4, no. 1: 101–10.

Bartlett, Tony, and Y. B. Malla. 1992. "Local Forest Management and Forest Policy in Nepal." *Journal of World Forest Resource Management* 6, no. 2: 99–116.

Bista, Dor B. 1991. *Fatalism and Development.* Hyderabad: Orient BlackSwan.

Blaikie, Piers, and Harold Brookfield. 1987. *Land Degradation and Society.* London: Methuen & Co. Ltd.

Carraher-Kang, Alexandra. 2019. "Patriarchal, Capitalist, and Colonialist Systems Lead to No Action at COP 25." *Cultural Survival,* December 23.

Corbera, Esteve, and Heike Schroeder. 2017. "REDD+ Crossroads Post Paris: Politics, Lessons and Interplays." *Forests* 8, no. 508: 1–11.

Damodaran, Vinita. 2006. "Indigenous Forests." In *Ecological Nationalisms,* edited by Gunnel Cederlof and K. Sivaramakrishnan. Seattle: University of Washington Press.

Dhungana, Sindhu, Mohan Poudel, and Trishna S. Bhandari, eds. 2018. *REDD+ in Nepal.* Kathmandu: REDD Implementation Centre, Ministry of Forests and Environment, Government of Nepal.

Eckholm, Eric. 1975. "The Deterioration of Mountain Environments." *Science* 189, no. 4205: 764–70.

Ervine, Kate. 2018. *Carbon.* Cambridge: Polity Press.

Gilani, Haris, Tomoko Yoshida, and John Innes. 2017. "A Collaborative Forest Management User Group's Perceptions and Expectations on REDD+ in Nepal." *Forest Policy and Economics* 80: 27–33.

Gilbertson, Tamra. 2017. "Carbon Pricing: A Critical Perspective for Community Resistance." Climate Justice Alliance and Indigenous Environmental Network, October. https://www.ienearth.org/wp-content/uploads/2017/11/Carbon-Pricing-A-Critical-Perspective-for-Community-Resistance-Online-Version.pdf.

Gilmour, Donald A. 1989. *Forest Resources and Indigenous Management in Nepal.* Working Paper No. 17. Honolulu: East-West Environment and Policy Institute, East-West Center.

Gulbrandsen, Lars H. 2012. "International Forest Politics: Intergovernmental Failure, Non-Governmental Success?" In *International Environmental Agreements,* edited by Steinar Andresen, Elin Lerum Boasson, and Geir Honneland. New York: Routledge.

Gurung, Jeannette, and Dibya Gurung. 2018. "Gender and REDD+ in Nepal." In *REDD+ in Nepal,* edited by Sindhu Dhungana, Mohan Poudel, and Trishna S. Bhandari. REDD Implementation Centre, Ministry of Forests and Environment, Government of Nepal. Kathmandu: Jagadamba Press.

Harvey, David. 2018. *Marx, Capital and the Madness of Economic Reason.* New York: Oxford University Press.

Joshi, Shangrila. 2019. *Climate Justice and REDD+: Voices of Community Forest Users.* White Paper. September 24. https://www.sylff.org/wp-content/uploads/2019/11/Joshi-2019-White-Paper.pdf

————. Forthcoming. "A Political Ecology of Neoliberal Climate Mitigation Policy in Nepal." Conference Proceedings 2019: The Annual Kathmandu Conference on Nepal and the Himalaya, Kathmandu.

Kane, Seth, Ahmad Dhiaulhaq, Lok M. Sapkota, and David Gritten. 2018. "Transforming Forest Landscape Conflicts: The Promises and Perils of Global Forest Management Initiatives Such as REDD+." *Forest and Society* 2, no. 1: 1–17.

Khatri, Dil, Thuy Thu Pham, Monica Di Gregorio, Rahul Karki, Naya Paudel, Maria Brockhaus, and Ramesh Bushal. 2016. "REDD+ Politics in the Media: A Case from Nepal." *Climate Change* 138: 309–23.

Manandar, Ugan, and Charlie Parker. 2018. "REDD+ Finance in Nepal." In *REDD+ in Nepal: Experiences from the REDD Readiness Phase*, edited by Sindhu Dhungana, Mohan Poudel, and Trishna S. Bhandari. REDD Implementation Centre, Ministry of Forests and Environment, Government of Nepal. Kathmandu: Jagadamba Press.

Mandal, Chandan. 2019. "Taking Back Forests from Its Community Users Is Curtailing Their Rights, Activists." *The Kathmandu Post*, May 27.

————. 2020. "Community Forest Users' Groups Decry Continuation of Heavy Taxes." *The Kathmandu Post*, June 4.

Mandal, Ram et al. 2013. "Evaluating Public Plantation and Community Planted Forests Under the CDM and REDD+ Mechanism for Carbon Stock in Nepal." *International Journal of Conservation Science* 4, no. 3: 347–56.

MoFE. 2018. *Nepal National REDD+ Strategy (2018–2022)*. Kathmandu: Ministry of Forests and Environment, Government of Nepal.

————. 2019a. *Submission of Voluntary National Report to United Nations Forum on Forests Secretariat*. Kathmandu: Ministry of Forests and Environment, Government of Nepal.

————. 2019b. *Environmental and Social Management Framework for the Proposed Emission Reduction Program Interventions in the Terai Arc Landscape*. Kathmandu: Government of Nepal, Ministry of Forests and Environment, REDD Implementation Centre, October.

Mohai, Paul, David Pellow, and J. Timmons Roberts. 2009. "Environmental Justice." *The Annual Review of Environment and Resources* 34: 405–30.

MoPE. 2016. "Intended Nationally Determined Contributions (INDC)." Communicated to the UNFCCC Secretariat, Government of Nepal, Ministry of Population and Environment, Kathmandu, February.

Mulmi, Amish R. 2017. "Why Did the British Not Colonize Nepal?" *The Record*, October 1.

Ojha, Hemant R., and Netra P. Timsina. 2008. "Community-Based Forest Management Programmes in Nepal: Lessons and Policy Implications." In *Communities, Forests and Governance*, edited by Hemant R. Ojha et al. New Delhi: Adroit Publishers.

Ojha, Hemant R., Netra P. Timsina, Chetan Kumar, Mani R. Banjade, and Brian Belcher, eds. 2008. *Communities, Forests and Governance*. New Delhi: Adroit Publishers.

Ostrom, Elinor. 1999a. *Self-Governance and Forest Resources*. Occasional Paper No. 20. Bogor: Center for International Forestry Research. www.umich.edu/~ifri/Publications/R991_4.pdf.

————. 1999b. "Coping with Tragedies of the Commons." *Annual Review of Political Science* 2: 493–535.

Paudel, Dinesh. 2016. "Re-Inventing the Commons: Community Forestry as Accumulation Without Dispossession in Nepal." *The Journal of Peasant Studies* 43, no. 5: 989–1009.

Pokharel, Bharat K., Peter Branney, Mike Nurse, and Yam B. Malla. 2008. "Community Forestry: Conserving Forests, Sustaining Livelihoods and Strengthening Democracy." In *Communities, Forests and Governance*, edited by Hemant R. Ojha et al. New Delhi: Adroit Publishers.

Poudel, Mohan, Eak Rana, and Richard Thwaites. 2018. "REDD+ and Community Forestry in Nepal." In *Community Forestry in Nepal*, edited by Richard Thwaites, Robert Fisher, and Mohan Poudel. London: Routledge.

Poudel, Mohan, Richard Thwaites, D. Race, and G. Ram Dahal. 2014. "REDD+ and Community Forestry: Implications for Local Communities and Forest Management – A Case Study From Nepal." *The International Forestry Review* 16, no. 1: 39–54.

Rai, Rajesh, Basant Pant, and Mani Nepal. 2018. "Costs and Benefits of Implementing REDD+ in Nepal." In *REDD+ in Nepal*, edited by Sindhu Dhungana, Mohan Poudel, and Trishna S. Bhandari. REDD Implementation Centre, Ministry of Forests and Environment, Government of Nepal. Kathmandu: Jagadamba Press.

Rangan, Haripriya. 2000. *Of Myths and Movements*. London: Verso.

Satyal, Poshendra. 2018. "Operationalising the Ideas of Justice in Forest Governance: An Analysis of Community Forestry and REDD+ Processes in Nepal." *The International Journal of Environmental, Cultural, Economic, and Social Sustainability: Annual Review* 13, no. 1: 1–19.

Satyal, Poshendra et al. 2019. "Representation and Participation in Formulating Nepal's REDD+ Approach." *Climate Policy* 19, no. 1: 8–22.

Shah, Saubhagya. 2018. *A Project of Memoreality*. Kathmandu: Himal Books.

Sharma, Bishnu, Priya Shyamsundar, Mani Nepal, Subhrendru Pattanayak, and Bhaskar Karky. 2017. "Costs, Cobenefits, and Community Responses to REDD+: A Case Study from Nepal." *Ecology and Society* 22, no. 2: 34.

Sherpa, Doma T. 2017. "Using the 3Es' Method to Evaluate REDD+ Project in Nepal." *Case Studies in the Environment* 1, no. 1: 1–5.

Sherpa, Pasang D., Tunga B. Rai, and Neil Dawson. 2018. "Indigenous Peoples' Engagement in REDD+ Process: Opportunities and Challenges in Nepal." In *REDD+ in Nepal*, edited by Sindhu Dhungana, Mohan Poudel, and Trishna S. Bhandari. REDD Implementation Centre, Ministry of Forests and Environment, Government of Nepal. Kathmandu: Jagadamba Press.

Shrestha, Binu. 2016. "Archaeological Findings Prove the Kasthamandap Was Built in 7th Century." *The Rising Nepal*, November 18.

Shrestha, Sujata, and Uttam B. Shrestha. 2017. "Beyond Money: Does REDD+ Payment Enhance Household's Participation in Forest Governance and Management in Nepal's Community Forests?" *Forest Policy and Economics* 80: 63–70.

Stumpf, Simon, Hermine Kleymann, and Kai Windhorst. 2018. "Results-Based Finance for REDD+: Approaches, Perspectives and Challenges." In *REDD+ in Nepal*, edited by Sindhu Dhungana, Mohan Poudel, and Trishna S. Bhandari. REDD Implementation Centre, Ministry of Forests and Environment, Government of Nepal. Kathmandu: Jagadamba Press.

Thwaites, Richard, Robert Fisher, and Mohan Poudel. 2018. "Community Forestry in Nepal: Origins and Issues." In *Community Forestry in Nepal*, edited by Richard Thwaites, Robert Fisher, and Mohan Poudel. London: Routledge.

United Nations. 2015. *Paris Agreement*, chapter XXVII, Environment, title 7d. Paris: United Nations, December 12.

Whelpton, John. 2005. *A History of Nepal*. Cambridge: Cambridge University Press.

World Bank. 2018. "People and Forests – a Sustainable Forest Management-Based Emission Reduction Program in the Terai Arc Landscape, Nepal." Forest Carbon Partnership Facility (FCPF) Carbon Fund Emission Reductions Program Document (ER-PD), May 23.

6 Learning from the Nepalese experience of taking back the forest commons

The climate crisis is in no uncertain terms a crisis of the commons. Not because humans are inherently selfish and cannot be trusted to manage them sustainably; not because there are simply too many humans, as Hardin (1968) hypothesized. Rather, we have a climate crisis because of the domination of a system of unregulated capitalism that regards the atmosphere and other resources as an open access resource (Foster 1999; Ostrom 1999b). Following Hardin's (1968) notion of the "tragedy of the commons", climate change is often understood to be a tragedy of the atmospheric commons, where the atmosphere is a global commons that carries out the essential ecological function of a sink for the absorption of GHGs (e.g. Ostrom 2014). A decade after publishing the 'Tragedy of the Commons', and being forced to respond to his critics, Hardin (1994) admitted to abusing the term 'commons' when he was speaking of open-access resources that are not regulated. Yet, his original set of prescriptions for how the 'tragedy of the commons' should be averted has proved to be hegemonic. Hardin (1968) had argued that the surest way to prevent a tragedy of the commons is to privatize it. After three decades of attempting to establish rules for governing the atmospheric commons by assigning mitigation responsibilities to the international community, we have now emerged with no such rules – with the exception of the soon-to-be-defunct Kyoto Protocol – but instead a commitment to continue with the market created for emission reductions introduced in the Kyoto Protocol (by the US negotiating team) as the ultimate global solution to the problem of too many greenhouse gases.

But the climate crisis is not just a crisis of the atmospheric commons. Rather, I suggest that the atmospheric commons is inherently tied to the forest commons, as demonstrated by the emergence of REDD+ as a climate solution (Ostrom 2014). It is also linked to fossil fuels, since burning these non-renewable resources is the major culprit (Griffin 2017); to water resources that are affected by a warming climate (IPCC 2014); and other commons. While the task of treating the atmosphere as a regulated commons has proven to be unachievable for the time-being, with the dissolution of the Kyoto Protocol, the ability of specific communities to govern specific resources pertaining to the atmospheric commons has not been diminished (Ostrom 2014). Chapter Five discussed the case of community forestry in Nepal where REDD+ interventions are being introduced due to the socially progressive forest governance structures therein. It remains to be seen whether the commons-oriented institution of community forestry in Nepal will eventually be eroded due to the pressures of commodification to suit carbon trading rationalities (Osborne 2015). If they do, the real tragedy of the atmospheric commons may be the emergence of the so-called neoliberal climate solutions to address climate change. Moments of crisis have historically offered a space of contestation between the oppositional forces of neoliberalization and commoning (Barrios 2016; Gibson-Graham, Cameron, and Healy 2013; Klein 2008; Solnit 2016). The outcomes of these contestations are neither fixed nor predictable, but are context-specific. It is therefore important to understand the specific ways in which a neoliberalizing intervention such as REDD+ may transform a commons-oriented forest governance structure such as community forestry in Nepal, while remaining open to the possibilities for affected individuals and communities to exert their agency within the constraints of neoliberal capitalism, given the understanding that "although hegemonic ideas are dominant, they are never total" (Shah 2018, 224). In undertaking a structural analysis of these entanglements, it is important to undertake an analysis of not only neoliberal capitalism but the particular ways in which this globalizing force interacts with prevailing structures of core-periphery inequality, state-society dynamics, particularly in the context of structures of gender and caste inequality, and tensions between the state and disenfranchised Indigenous groups.

Hariyo Ban Kasko Dhan? A Structural analysis of community forestry in Nepal

The Nepali government has had the slogan *Hariyo Ban Nepalko Dhan* – meaning, Green Forests, Nepal's Wealth – for as long as I can remember. But who exactly is represented by Nepal here? And how is wealth understood or measured, and by whom? Nepal's population is gendered and racialized to create particular social hierarchies that are entrenched. This population is further segregated into metropolitan cities such as Kathmandu that form the core of the economy with peripheries that have until recently had a relationship of dependency to the core, in much the same way as its own dependency towards bilateral donors in Asia and the Global North. The Nepali state has long had an extractive relationship with its citizens that enables a view of *Nepal's wealth* to mean the *wealth of the Nepali state*. Whether to consider forest wealth in economically reductionist terms of financial wealth or in more holistic ways including ecological vibrancy and social-ecological capability is an ongoing question (Sheppard et al. 2009). With the entanglement with REDD+, these questions broaden the scope of 'forest wealth' beyond forest ecology, and beyond Nepal's borders, to encompass the global atmospheric commons. Consequently, when CFUGs in Nepal do the work of keeping forests intact, they are now not only creating 'wealth' for their communities, villages, and country, they are also doing the work of sequestering carbon and contributing to averting climate collapse, while enabling the profit accumulation of those whose GHG mitigation commitments are being subsidized. If this work is being done in the pretext of an international transaction – carbon sequestered in exchange for financing forest governance – what would constitute an equal exchange or fair trade? Who gets to decide?

> They tell us to conserve forests. Is it possible that they are making us into servants? Their own environment is polluted. They tell us not to use our resources. Could it be that that is why we aren't more developed? Do they always want us to be dependent on them? Aren't the developed countries the ones who are polluting day after day?
>
> (Bahun/Chhetri Female, Chaturmukhi CFUG)

> In Europe they cut down all the tees, they developed. They ask us not to cut trees. People say it may be because they want to prevent us from developing.
>
> (Male, Jaldevi CFUG)

> Capitalist countries don't change anything. They would rather buy carbon from countries like ours. Give some compensation, some money, but change nothing in their own. Their own privileges are not affected. Only our privileges are affected. Our development is curtailed. The NGOs that come here to tell us don't do this or that with our forests, the developed countries have done it all, dozers, airport, highways, military. And they tell us what to do and not do. US is giving 50 million dollars for an electricity project, but there are conditionalities. And now this carbon trade. In capitalism, the people at the top enjoy all the privileges . . .

> Europe and US got wealthy by colonizing South America and Africa, their iron, wood, etc. and now they are being asked for reparations. India with Britain. Similarly in Chitwan rhinos, tigers, and bears have thrived at the expense of your ancestors. You are still suffering. Tourists from Kathmandu and elsewhere who come here, what do they owe you? Now with carbon stock, they don't want to count the carbon stock of the last 50–60 years, only new carbon. These issues have to be raised. The historical contributions have to be accounted for. Our country has not been exploited as much but still there are debts.
>
> (Bahun/Chhetri male, Gorkha FECOFUN, August 2019)

Moving beyond a grand theory of neoliberal capitalism

Political ecologists strive to understand the root causes – the structural explanations – underlying the socio-ecological changes they study. From the beginning, the discipline has been shaped by Marxist political economy; therefore, an analysis of capitalism and how it shapes the behavior of land managers and other resource users has been the bedrock of political ecological structural analysis (Bryant 2001; Robbins 2004). As such, the roots of the ecological crisis are traced to the political economy of advanced capitalist societies that are ideologically wedded to infinite economic growth. Capitalism is seen as inherently exploitative towards nature and the laboring classes, and a regulated, ethical capitalism is seen as a contradiction in terms (Foster 1999). Harvey (2018) has described capitalism as a system of accumulation by dispossession, where profit accumulation for the capitalist class necessarily involves dispossession for members of the working class, who may be disenfranchised from their resource base or alienated from their labor. As resources and labor are depleted in certain regions, the resourceful capitalist manages to find new terrains to exploit, which is the notion of the spatial fix – a temporary solution to averting the inevitable crisis of capitalism (ibid). The recent announcement by the Trump administration that the United States would seek to explore the possibility of colonizing the moon and Mars, with an eye to promote the mining of their resources by private entities perhaps, signals the ultimate spatial fix. Its refusal to acknowledge outer space as a global commons is a telling reminder of our failure with the Kyoto Protocol.

Capitalism is not a self-sufficient system, but is connected to the structures of colonialism, racism, and patriarchy. Federici (2004) has noted that it was in the transition from feudalism to capitalism in Europe during the 15th–17th centuries – and the enclosures of the commons – that women were disenfranchised from their power over their bodies, labor, and from the commons. Merchant (1983) showed that the logic of extractivism that accompanies capitalist accumulation is enabled by a misogynistic view of both women and nature that was prevalent in Baconian science. Both Merchant (1983) and Foster (1999) noted the role played by science in facilitating capitalist exploitation. Marxist political economy has tended to see colonialism as just an extension of the capitalist project (e.g. Blaikie and Brookfield 1987), but others object to such a hierarchy of explanations (Estes and Dunbar-Ortiz 2020) and problematize such Eurocentrism as a form of coloniality (Grosfoguel 2011). Yet others have asserted that an exclusive focus on a Marxist class analysis is "incomplete and, in a certain sense, parochial – limited, to a great extent, to the European experience of the nineteenth century" (Laclau 1990, 130) and problematized the presumption of Eurocentric depictions of capitalism and class struggle as the *only* true form of it (Blaut 1993). Indeed "there is therefore no 'capitalism', but rather different forms of capitalist relations which form part of highly diverse structural complexes" (Laclau 1990, 26). Rather than examining forms of accumulation by dispossession in the context of the climate crisis strictly within the paradigm of neoliberal capitalism, they should be understood in a broader context of the hegemony of patriarchal-modernist-capitalist-colonialism (Grosfoguel 2011). Likewise, counterhegemonic responses should be sought not only with a singular vision of socialism but with a

> plurality of social agents and of their struggles. Thus the field of social conflict is extended, rather than being concentrated in a "privileged agent"

of socialist change. This also means that the extension and radicalization of democratic struggles does not have a final point of arrival in the achievement of a fully liberated society. There will always be antagonisms, struggles, and partial opaqueness of the social.

(Laclau 1990, 130)

Although Marxist political ecology offers compelling explanations for the climate crisis, much of it is inadequate in explaining socio-environmental change in the Nepalese context. Some of the earliest work in political ecology is in fact premised on studies of soil erosion conducted in Nepal that sought to rescue the peasant farmer from the clutches of neo-Malthusian theories similar to Hardin's (Blaikie 1980; Blaikie and Brookfield 1987), but in the process were themselves prone to a different kind of oversimplification of Nepal's political economy – reducing Nepali society to simply a periphery of Western capitalism, colonialism, and imperialism, while failing to understand the nuanced complexities of caste structure and ethnic politics in the region, which cannot be understood by simply using Western constructs or even India's experience of caste and colonialism (Bista 1991; Mishra 1980). While the incorporation of Nepal's community forests into global carbon trading circuits makes an understanding of neoliberal capitalism critical for Nepal, it is equally important for political ecologists and other scholars to get the pre-REDD+ CF context right. In this sense, Blaikie and Brookfield's (1987) "chain of explanation" needs to expand beyond a theory of European colonialism and capitalism, and unpack regional variants of both, and how they have produced Nepal's peculiar patriarchal caste structure, the Nepali state's tenuous relationship with its Indigenous Nationalities, and Nepal's bilateral relations with its neighbors and Western states. Further, as Bumpus and Liverman (2010, 212) remind us:

A critical political ecology analysis that reflects the ways in which carbon finance is resisted or 'reworked' into more positive local impacts is an important practical and theoretical endeavour. . . . Carbon offsets must be conceived as relational: only by analysing the relationships that exist between the carbon emitter in the North and the carbon reducer in the South can the environment-development implications for offsets are understood.

Accumulation with and without dispossession

Paudel (2016) has surmised that Nepal's community forestry is one of the rare exceptions to the theory of accumulation by dispossession, arguing that accumulation *without* dispossession can be witnessed here. The Government of Nepal generates revenues through taxes on timber sales. Community forestry has reportedly generated more than USD 50 million annually from the sale of timber and non-timber forest products in recent years (Dhungana, Poudel, and Bhandari 2018). Although CFUG members do not own their means of production, the community forests, they are also not displaced from them even as the timber in these forests may be sold for profit. Moreover, in addition to being able to continue to derive subsistence benefits from the forest – based on an annual allowable harvest for household use based on physiographic region (Rai, Pant, and Nepal 2018) – CFUG members may make income from the sale of forest products (Paudel 2016). According to the 2014 Community Forestry Guidelines, CFUGs have to invest their earnings on community development, including capacity-building and infrastructure projects (40 percent), livelihood enhancement (35 percent), and forest conservation (25 percent) (Mandal 2019). Any timber (e.g. *Sal* Shorea robusta, *Khair* Acacia catechu) sold outside CFUGs is taxed, with the revenues going to various levels of government, including 15 percent to the government's central treasury (Pokharel et al. 2008). In 2008, taxes on the sale of forest products from community forestry were estimated at a total of USD 8.5 million (Manandar and Parker 2018). Since 2015, additional taxes have been proposed at three layers of government – local, province, and federal – to add up to 65–90 percent tax obligations for CFUGs based on province. Following concerted protests waged by CFUGs and FECOFUN, the federal government retracted the 15 percent federal tax, but CFUGs are still unhappy with these new developments (Mandal 2020). It appears that the tax hikes are motivated by the need to provide co-financing as per the GoN's ERPA with the World Bank, but it is unclear to me what the relationship between the two is and how it will unfold. While it is noteworthy that a process of capitalist accumulation can co-exist with community autonomy over forest resources (Paudel 2016), it is nevertheless worthwhile to ask who benefits from the particular ways in which community forestry is structured and whose reproductive labor sustains this structure.

Forestry has served as a source of revenue for the state's ruling elites since it was officially introduced during the Rana regime (1846–1951) (Ojha and Timsina 2008; Pokharel et al. 2008). Consequently when it was deemed more profitable for the state to generate tax revenues from agriculture through feudal-aristocratic groups, deforestation occurred in the early 1900s in the Terai forests to create land for agriculture. Even after CF was introduced, the government has handed over forests deemed economically less valuable to CFUGs, keeping the Terai forests containing much of the profitable *Sal* forests within its own domain: merely 3 percent of Terai forests had been handed over to CFUGs, and only 7 percent of CFUGs were in the Terai (Pokharel et al. 2008). There is also an economic logic for the government to let CFUGs manage forests instead of doing it on its own since it is cheaper to do so (Rai, Pant, and Nepal 2018). When something is cheap, it is usually because an externality has not been accounted for (Stanford 2015). In a colonial-capitalist context, cheapness is an outcome of exploitation of labor and resources (Acha 2019). When reviewing the operating budget for community forests, it is clear that the cost-saving owes to unpaid labor costs: CFUG labor constitutes 70 percent of the budget, in addition to cash contributions from dues-paying members, 25 percent. The labor value invested in CFUGs annually was estimated to be about USD 12 million (Pokharel et al. 2008). Since CFUG revenues sponsor community development projects, including drinking water, school buildings, temples, trails, telephone service, and so forth, they further ease the burden on the government in attending to these responsibilities. Yet, until recently their contributions to the nation's economy had not been recognized (Pokharel et al. 2008; MoFE 2019). This invisibilization renders it akin to feminized reproductive labor (Waring 1990). Seen in this way, one could argue that CFUGs are not fairly compensated for their reproductive labor costs in sustaining the forests that hold such economic value. This undervaluing works at multiple scales – gender, community, national, and now with REDD+ at the international scale as well. Heterogeneity and hierarchies at the community level enable male dominance and elite capture, which is reflected in terms of wealth, caste, and Indigeneity/ethnicity.

In the aphorism *Hariyo Ban Nepalko Dhan*, it is starting to be clear for whom forest wealth has traditionally been generated and by whom. It is the free or cheap labor of the forest user that is behind the sustenance of Nepal's community forests, now valued for contributing to preventing climate collapse. *Nepalko Hariyo Ban* is now *Sansarko Dhan* – the world's wealth. *If all forest stewards had knowledge of this, how would they react? How should their contributions be valued, monetarily and otherwise?* Many of the CFUG members I interviewed shared a keen sense of injustice, how the problem of climate change – that was resulting in erratic rainfall, hotter weather, floods, and landslides – was created by developed countries with their polluting factories, cars, and rampant development patterns. Nepal on the other hand was seen as a haven for nature and forests, and among those countries that were absorbing their carbon. The work of conserving forests was done by forest users risking their lives due to the risks from wildlife protected in nearby national parks. From the vantage point of CFUG leadership, the irony of the injustice was that the ordinary forest user is unaware that they are doing the work of sequestering carbon. On top of this unequal exchange, developed countries and the Nepali state were seen to benefit from carbon trade as well. Without being duly informed by the state about the specifics and the complexities of carbon financing, CFUG leaders articulated their fears, hopes, and concerns regarding CF and Nepal's involvement with REDD+.

> Roaming the forests is high-risk work, with snakes, tigers, rhinos threatening our lives. The work that the police and the military have not been able to do [CF] has done, in recovering forests. The government didn't just gift [CF] to us, we demanded it. Now if the government gives us no avenue for income but instead wants to tax us, it is not fair.
>
> (Adibasi/Janajati female, Chaturmukhi CFUG)

> In the long term it seems there will be negative effects. I am no expert. From my point of view, if we enter carbon trade we are only going to be the *ban pale* [foot soldiers] of community forests. We conserve. The international community continues to pollute. If their well-being depends on

us, we may have to continue doing this work. The developed world is not going to stop developing. Factories, railways, pollution will only go up. Might we end up being security guards? Kailali has 610 community forests – 30 have adopted [SF] – two collaborative forests and one private. Kailali has 64 percent of Nepal's forests. 62 percent of Kailali is forested. 5,939 hectares are community forest. Who is it for is the question. So far it has been for ourselves. Now if we do it for money for somebody else, what if they say we can't go in to our forests? But even if you place constraints on people, they will take what they need. There may be widespread conflict if the policies are unfair or unclear. There needs to be clear information about policy all the way to the bottom. Even we who are position holders in [CF] don't know anything about this, let alone those who are on the ground doing the actual work of forest protection. If information cannot get to them, this cannot be sustainable. Everyone needs to understand national and international policy related to this . . . is REDD+ fair or unfair we don't know . . . will there be carbon trade? I don't know, and neither does anyone in my community. Maybe experts do but not anyone in the [CF] context. I think we should know every single thing about it. A 10-page document should clarify everything about REDD+ and how it works. It should be integrated into our [CF] action plan. They cannot just bring a book and ask us to sign.

(Dalit male, forum participant from Kailali)

Carbon is a business. There is no guarantee of a profit, nor that the people in the most poor and marginalized categories will see an increase in income, resources, and economic progress. Moreover, if the carbon measurements and inventory work are done by experts, which is what has happened already, any financial gains will leave the community to pay for their salaries in US dollars. Their work may be specialized and may be necessary to document carbon increase, but the real work of conserving the forest is done by the local non-expert. That person has to understand and get the benefits. . . . The person who is saving needs to understand what they are saving, why am I saving it, and for whom. Who benefits? Unless this person knows what they are doing and why, the policy level people's grand plans won't be successful. When there is a fire it is they who go in to protect the forest. REDD+ eventually depends on forests and it is they who protect them.

(Bahun/Chhetri male, Amrit Dharapani CFUG)

Currently the Nepalese government has unilaterally enlisted the involvement of CFUGs in facilitating carbon sequestration-based financing, with a formalistic consultation at most (Satyal et al. 2019). Since carbon is an elusive commodity, Rai, Pant, and Nepal (2018, 92) have urged that

any facilitation of carbon trade between local communities and international buyers of carbon should be based on a sound understanding of a) existing carbon stocks in a particular forest; and b) the opportunity costs of conserving these forests for sequestering carbon, i.e., the costs communities will incur when they conserve forests for carbon purposes.

There are numerous unresolved questions regarding how stakeholders of forest governance at multiple levels will be meaningfully involved in the prioritization of competing forest values and uses, or negotiations over equitable benefit-sharing of carbon finance. Seeking to establish a monetary price on the value of benefits such as biodiversity and habitat conservation has always been a near-impossible task (Stanford 2015), and would be made more so, with the multiplicity of land tenure regimes in Nepal, although studies have produced conservative estimates of how ecosystem services of Nepal's forests might be valued (Rai, Pant, and Nepal 2018). To be fair, most of the ecosystem services are enjoyed by the CFUG members themselves, whether or not they are compensated for them. For instance, the work they do to prevent forest fires by employing forest guards from the local community helps with not only carbon sequestration but their own socio-ecological communities and livelihoods. Economic determinism in cost-benefit calculations can conceal the intangible benefits derived from the devolution of forest governance to local communities:

> Community forestry, through creation of experiences with local decision-making, institution-building, leadership, conflict resolution, common

Figure 6.1 Premises of the Amrit Dharapani CFUG office, Chitwan, August 2017

resource management and forest-based enterprise, is therefore fertile ground for rural people to develop their own skills and attitudes in these and for contributing to wider society outside the forestry sector. It is not surprising that many VDC office-bearers have built their reputations from first being involved in CFUGs. This has allowed them to gain the respect and recognition of their neighbors and to become more actively involved in local governance.

(Pokharel et al. 2008, 77)

The struggle over the commons: power struggles with the state

Cole and Foster (2000) described the struggle for environmental justice as a transformative process where those in pursuit of justice are empowered by the process of changing the power dynamic inherent in the status quo. The struggle for the forest commons in Nepal and the consequent movement from state-centric to participatory governance of forests can be seen in a similar light. It is a struggle that was not complete even when community forestry legislation was ushered in in 1993, and following a tumultuous history including several reversals, it is a struggle that is still ongoing. As Ojha and Timsina (2008, 215) observed, "land and forest resources continue to remain the principal means to sustain the nation-state, as well as the linchpin for repeated political upheavals". Community forestry has been referred to as an iterative process that has enabled a "dialogical interaction between practice and policy", where modest policy reform within bureaucratic spaces is followed by innovative practice in the forests, which feeds back to radical reform in policy spaces (ibid, 225; Pokharel et al. 2008). This process has not been easy to pull off. As noted, a progressive forest policy had been crafted in 1953 that included provisions for rural forest-dependent communities to meet their subsistence needs (Pokharel et al. 2008) but had been shelved in the ensuing Forest Act of 1957. Pokharel et al. (2008, 57) therefore noted that the "history of CF in Nepal subsequently became one of a struggle between communities who were gradually trying to acquire or re-acquire rights and control over forests and a largely resistant state (in the form of the Forest Department)".

The presence of foreign actors has long played the role of a double-edged sword in these state-society struggles. To understand this more fully, we have to go back to the point of origin of the Nepali state in the 18th century. The impetus for the 'unification' of the erstwhile *baise* and *chaubise rajyas* – the conglomerations of 22 and 24 principalities – at the foot-hills of the Himalaya was the approaching threat of the British (Bista 1991). The visionary individual credited for thwarting British colonization of present-day Nepal was Prithvi Narayan Shah, a young King of Gorkha, one of the 24 principalities, known to be ruled by a Thakuri lineage (including Ram Shah) that claimed descent from the Rajputs of the Rajasthan region of current-day India, who had escaped persecution from Mughal invaders during the 14th and 16th centuries and sought refuge in the hills to the North (Whelpton 2005). In the process of creating a 'unified Nepal' many Indigenous communities and states – primarily of Khas and Kirat lineage – would come under the brutal rulership of a Nepali-speaking Hindu King. Kathmandu would become the new capital of the Nepali state formed in 1768–9 after the three Newar Kingdoms in the Kathmandu Valley – Lalitpur, Basantapur, and Bhaktapur – were defeated in a 25-year war (Allen 2015; Bista 1991; Levy 2019). The Kathmandu Valley was referred to by the Indigenous Newar therein as Nepa (Birkenholtz 2018). Nepal's 'unification' was therefore itself a process of appropriation of Newar identity and culture, and colonization of the land and resources – both natural and human – of the erstwhile nation-states while establishing a Hindu hegemony with its caste, class, and gender hierarchies, and invisibilization of Indigeneity (Bista 1991; Regmi 1967). The counterhegemonic struggles for recognition, equality, and autonomy by the oppressed caste (Dalits) and Indigenous Nationalities (Adibasi/Janajati) have been ongoing, as reflected in protests against the 2019 Guthi Bill and the taking of land in Khokana for construction of a 'Fast-track road' that represents new iterations of threats to Newar autonomy over agricultural land and institutions in the name of development (Manandhar 2018; Shrestha 2019); and contestations against the new wave of Sanskritization in school curriculum by the Nepal Federation of Indigenous Nationalities (NEFIN) (Ghimire 2020; Online Khabar 2020).

Contestations over the 2019 Forest Bill, now passed into the 2019 Forest Act, do not follow Dalit/Adibasi/Janajati fracture lines but rather reflect ongoing struggles between diverse forest users in the peripheries against rent-extracting tendencies of the Nepali state forest bureaucracies located in Kathmandu – comprised mostly of male caste Bahuns of Khas/Arya lineage – as the political economic core. With new forest regulations, forest users are seemingly enlisted in a financing scheme for REDD+ investment – their CFUG contributions are counted as equity investments and private sector co-financing (Manandar and Parker 2018). With CFUGs imagined as private investors – presumably without their consent – the identity of CF as a commons is possibly jeopardized; now each individual forest commons (CFUG) becomes a private entity – a co-investor – in the newly commodified atmospheric commons. This dance between forests seen as private and forests seen as commons goes back to at least 1957, when the Private Forest Nationalization Act facilitated the seizure of forests privately owned by those in the immediate orbit of the ruling class, specifically the Rana courtiers of the Shah Kingdom who had seized power to exert autocratic rule in Nepal from 1846–1951. Their ousting was facilitated by allies in a newly independent post-colonial state of India. Just as the Rana regime was coming to an end, the post–World War II international development agenda had been supporting a strong welfare state. Inspired by developments in post-colonial India, Nepal was at this time witnessing its own movement towards democratization, culminating in a multi-party elected parliament to work alongside the Monarchy in 1959. Nationalization of forests held privately by the Rana regime and their adherents was made possible by this momentum. Once private forests were nationalized, the Nepali state enhanced its monopoly over forests. The Ministry of Forests was created in 1959, followed by the Timber Corporation of Nepal in 1961, and the Fuelwood corporation in 1966, alongside rhino and wildlife conservation programs (Ojha et al. 2008). The Forest Acts of 1961 and 1967 were introduced, making "traditional community activities illegal" (Pokharel et al. 2008, 56). In the 1960s, King Mahendra seized power, dismantling the elected parliament and replacing it with the Panchayat system of local governance (Ojha et al. 2008). Nationalization of forests in this political environment

> replaced a complex, although well-understood, system of de facto local ownership, usufruct rights and use-patterns (some of which were indeed elitist and inequitable) with a monolithic and unwieldly governance structure, where control over all forests became vested in the hands of a distant and inevitably resource-strapped government.
>
> (Pokharel et al. 2008, 56)

Although owned by the ruling class, much of the forested land in the Terai that was seized by the state had been occupied and used by the Indigenous Tharu, Bote, Majhi, and Chepang, among others. Nationalization of forests did not empower them, but rather, criminalized their activities. With their de facto tenure and rights undermined, local people were disincentivized from acting as forest stewards. Consequently, rampant deforestation followed due to forest encroachment and unregulated extractivism, facilitated by a tense relationship between the local people and the forest agency (Ojha et al. 2008). But nationalization eventually paved the way for decentralization of the power seized by the state – albeit in gradual steps – spurred by a series of bottom-up people's movements, with foreign interventions serving as catalysts, and peppered with periodic reversals of progressive reforms (Ojha et al. 2008; Pokharel et al. 2008). The first tentative step in decentralization of forests was the provision of handing over the management of degraded forests to local governance bodies called Panchayats – often comprised of the traditional rural elite in terms of wealth, caste, gender, and education – as per the Panchayat Forest Rules in 1978 (Ojha and Timsina 2008; Pokharel et al. 2008, 56). This 'Panchayat-centered decentralization' would evolve into a more genuine devolution to locally formulated forest communities, bolstered by the Western environmental agenda during the 1970s and 1980s: "Since donor agencies were effectively providing most of the development funds required by the government to implement the community forestry programme, they also became highly influential in terms of the policy and legislative framework to enable these approaches to succeed" (Pokharel et al. 2008, 59). A dialectic of West-driven progressive agendas and the bottom-up democracy movement of 1990 eventually solidified the notion that those 'forest users' who are affected by decisions made about forests they traditionally use should be involved in decision-making (ibid). Such participatory justice processes have been identified as crucial to environmental justice struggles. The formation of community forestry in Nepal therefore offers a successful example of an environmental justice struggle from the perspective of state–society relations; from the point of view of gender/Dalit/Adibasi/Janajati rights, the struggle is still at a nascent stage.

Reversals and power hoarding by the state

But as evidenced in EJ gains in other places such as the United States, hard-won gains may not be complete, and even incomplete gains can be reversed. Due to the liberatory effect of CFUG formation on local power geometries, reversals have significant political implications (Ojha and Timsina 2008). The 1993 Forest Act recognized CFUGs as

> legal, autonomous and corporate bodies having full power, authority and responsibility to protect, manage and utilise forest and other resources as per the decisions taken by their assemblies and according to their self-prepared constitutions and operational plans (with minimal scope for inter-ference from the state forestry agency).
>
> (Pokharel et al. 2008, 60)

Yet, even as the government officially 'handed over' forests to CFUGs recognized as "autonomous and perpetually self-governed organizations" (Ojha et al. 2008, 29) that could formulate rules for resource use, and mete out sanctions against rule violators – essentially using the forests within their purview as a commons (Ostrom 1999a, 1999b) – it retained ownership of the land and the power to make ultimate decisions concerning forests, including determining sustainable harvest levels and sales of timber (Ojha and Timsina 2008). The state didn't offer to share power out of generosity, though; it was wrested away through struggle. There remained powerful entities within the forest bureaucracy of Nepal who were against devolution, and they succeeded in unilaterally implementing several reversals of the radical reforms enacted by the 1993 Forest Act, gradually eroding and undermining CFUG authority and autonomy, and weakening the deliberative process. Since 2006, CFUG executive committees are ultimately answerable to the state forest bureaucracy rather than to CFUG members, since DFOs were empowered to mete out punishments to CFUG leaders (Pokharel et al. 2008).

Even as relatively degraded forests were handed over to CFUGs, the most lucrative forests of the Terai have been held by the state. Since 2000 new CFUGs have not been created in the Terai, with Collaborative Forests being the preferred CBFM as per the Terai Forest Policy – similar to India's Joint Forest Management, these impart greater authority than CF to the state and use scientific forestry practices more widely (Ojha et al. 2008; Rai et al. 2018).

This shift to collaborative forests has been contentious and has raised concerns from civil society about corruption. As a CFUG member shared,

> Collaborative forestry is an excuse to get Sal to the South of the border to India. Elected officials facilitate such *taskari* [smuggling]. Nepal *sarkaar* [government] looks the other way . . . We call [CF] ours but we haven't been able to use its resources . . . We are traumatized by collaborative forestry.
>
> (Adibasi/Janajati male, forum participant from Parsa)

The rationale given by the state was that the Terai forests were "not only property of local communities . . . but also the property of the nation, the larger public; and second, the forest should also be retained for its protective function apart from community or public utilization" (Ojha et al. 2008, 44). This is a politics of scale on part of the state, appealing to the notion of local forests as global heritage as a way to control high value forests. Further, they did so "by appealing to notions of equity" and problematizing the geographic exclusion of 'distant users' near the Southern border as per CF rules. These distant rulers are also Indigenous to the region, and they have been disenfranchised from forest access and displaced from their ancestral territories due to the establishment of national parks. Whether the 2000 Terai Forest Policy actually helps those distant users is uncertain. As of 2020, the Chepang were facing displacement due to their homes allegedly being burnt by park authorities due to perceived encroachment (Paudel 2020). Such equity arguments have nonetheless been employed to push for more collaborative forests in lieu of new community forests in the Terai (Ojha and Timsina 2008).

The state has employed many tactics to hold on to power over the forest commons. Ojha and Timsina (2008, 29) go so far as to suggest that the semblance of devolution was in fact a strategy on part of the repressive Panchayat government to co-opt potential dissent and to maintain hegemony. Pokharel et al. (2008, 65) refer to the process as "centralized decentralisation" since the participatory process has become largely ornamental. In this light it appears reasonable to suggest that the elevated and possibly romanticized governance powers of local forest users serve as a façade for the subtle and insidious processes through which the state concentrates its power over resources. The state has also hoarded power and exercised reversals through "technocratic domination in policy making, programme planning and implementation" (Ojha et al. 2008, 30), "excessive bureaucratic procrastination, imposition of 'guidelines', and ad hoc individual interpretation of grey areas" particularly since the 2000s (Pokharel et al. 2008, 63). These techniques concentrate power in the hands of experts and create the circumstances that render forest users tangential to the deliberations. One of the ways in which this is being done is by promoting scientific forestry (SF) – an approach to forestry that is skewed towards optimizing timber harvests – in the name of sustainable forest management. SF is currently a heavily debated issue in CFUGs across Nepal. During my fieldwork, CFUG members in Chitwan who had become tired of waiting for new REDD+ monies felt that a community forest in Korak was chosen as a model community forest and was receiving funds from the DFO because they had adopted SF (Interviews with several CFUG members in Shaktikhor, July 2019). SF is compatible not only with maximizing optimal yields for harvest but also for facilitating carbon measurements or "rendering carbon legible" (Osborne 2015). Agrawal (2005, 34–35) has argued that SF has been an enduring legacy of British colonial rule in forests in India; its colonizing effect works as a tool of governmentality no matter who wields it:

> Advocates of decentralized as well as centralized government of forests thus wield statistical arguments in the defense of their goals. It would seem that statistical facts can be yoked to multiple roles. But it seems equally evident that once precise, statistical, generalizing arguments are invoked in the service of policy it is difficult to counter them with vague, descriptive, anecdoctal evidence. It is in this characteristic of statistical representations – their capacity to displace nonnumericized arguments and advocacy – that their colonizing effects are to be found.

There are emerging tensions concerning whether CFUGs should embrace SF. While a few people viewed it favorably from the point of view of enhanced productivity and realized profits, most saw it as an externally driven agenda, an unwanted imposition from the state, one that threatened CF autonomy and identity, and imposed a heavy economic burden on CFUGs due to the need to hire technical experts for the required Environmental Impact Assessment. Many who viewed it unfavorably shared a strong visceral reaction to the image of a landscape where a whole swathe of trees had been felled as per the SF practice of harvesting by rotation, preferring instead, the practice of selective harvesting that is traditionally practiced.

> *Baigyanik ban* [SF] may be bad in the long run. In the name of [SF] everything is being cut down. If we don't have forests, how will we get carbon money? . . . Carbon management is okay but how much carbon is there? We need to know. Then the carbon trade needs to be just if it is to be sustainable, or else there will be conflict. We have to think long-term.
>
> (Adibasi/Janajati male, forum participant from Parsa)

We are divided here about [SF]. If the trees are cut down how will you have carbon trading? My [CF] went from having seven crores to five lakhs. A lot of the expenses were unnecessary. Forests that were preserved since 40–50 years are now being cut down. There will be no *Sal* left in the Terai belt. This is the government's policy. We will find ourselves saying "once upon a time we had *Sal Jangal*". . . . People at the top levels of forest bureaucracy, parliament, don't understand the issues. These new Forest

Acts. Ministers say different things in private and public. [SF] is really bad. If everything is gone, how will we trade carbon? We fear REDD+ will become like national parks. We used to get grass, wood, but no more. Our rights in [CF] are eroding, things are looking bad.

(Dalit Female, forum participant from Kapilvastu)

There are discussions to expand the national park, not just in our region but also in Parsa and Bara. Two delegations went to the forest ministry in Kathmandu to protest. Our local leaders cooperated without consulting with us but there is nothing that we cannot achieve if we decide to protest. We have protested tax hikes in the past. Again the DFO came to lure us into adopting [SF]. They said they would give us 25 lakhs. We are not going to be blinded by that kind of money at the cost of our future. We have to keep making our voice heard, protest what is not right. Not accept everything. If the government keeps encroaching on our rights, we shouldn't be silent.

(Dalit female, forum participant from Rautahat)

It is perhaps paradoxical that while on one hand REDD+ introduces social safeguards ostensibly as a conditionality for enhancing intracommunity equity particularly focused on gender and Indigenous rights, it is simultaneously creating the conditions that may undermine CFUGs' self-governance, autonomy, and leadership won after long and hard battles facilitated by an earlier wave of a Western environmental agenda. Although it is difficult to indisputably ascertain cause and effect, the ushering in of a wide range of laws, Acts, and policies pertaining to the 'sustainable' management of forests has coincided spectacularly with REDD+ preparations and programming. With REDD+ financing on the horizon and a slew of forest policies signaling reversal of gains made in community forestry, it appears that the forest commons are in the process of becoming privatized in multiple ways, including the incorporation of CFUGs as private co-financers, the entanglement of forest carbon with tradable carbon credits – euphemistically labelled as carbon financing – and the associated push towards SF. The outcome may be to negate the achievements made towards a progressive model of community forestry governance, in which case the legitimacy of REDD+ would be rightly called into question. Therefore the state as well as multi-lateral and bilateral sponsors of REDD+ would do well to heed the desires of forest users for their priorities to be valued and for more and meaningful deliberations involving local stakeholders in pivotal decision-making.

The bitter truth is that if not for the 1995 policy, we wouldn't have any forests. They were bare. Strict policies don't work. When forest guards were responsible, forests were dwindling, there was illegal smuggling. They didn't care. If not for [CF], the state of our forests would have been unimaginable. The public ushered in [CF] with the help of the nation, forest

Figure 6.2 Three-day forum on REDD+ and Climate Justice organized by author, Pokhara, August 2019

experts, and the international community. Forests were restored. But that is not enough. Those who are the protectors, who live close to the forests, their rights must be protected. We cannot be asked to do the work and someone else make the policies. Take, for example, the triple taxes to local, provincial, and federal agencies. In this manner, the forests will not be protected. Forest-dependent people are starting to ask why should they protect forests. . . . Carbon trading is not bad, but the problems always arise with the economic side of things. That is what creates conflict and leads to failure. REDD+, INGOs will get blamed, so from the start, the policies should be clear. There should be no uncertainties. The ones working the hardest should see benefits.

(Adibasi/Janajati male, forum participant from Banke)

Patriarchy

I didn't like to go anywhere. I used to stay within the home. What would I do outside in the world? We are women, the thinking used to be that we shouldn't go out. One of the *dais* [older brothers; generic term for males in the community] encouraged me to get involved, said I might learn something. I said "okay", and it turns out you do learn many things. We didn't used to go anywhere. We would stay in the home, rear children. Tend to goats, buffalo. We are surrounded by the old thinking that women shouldn't get ahead. No matter what anyone says, this is still the way people think, even in more educated homes. In the majority of Nepal where education is still largely absent, the old way of thinking is still there. Even so, it would have been alright if the focus was on good things, but negative things affect a person. For instance, as per tradition, if you are a girl, you cannot interact with males outside. This mentality is imparted from a young age, so you don't. Even if I did walk with males, how do I know how to conduct myself? How should I speak? How are others judging me? I don't feel capable. I experience *khinnatabodh* [self-consciousness and dejection]. But if someone says, "no you have to go out there", and gives that kind of encouragement, then you can have confidence. I wish more women could develop that confidence. In our Nepal, there are countless women who have been *peedit* [oppressed] like me. When I go out to do my forest work, I meet other women who understand these things, but there are still so many women who cannot even say their own name. If someone says "why don't you try becoming a member, you might learn some things", often the answer is "no, I am not educated, I don't know anything, I don't know how to speak, why go?" So even if women's participation should be 50% by law – I don't know if it has been imple-mented everywhere or not, the government raised it from 33% to 50% – but if they could reach more women who are left behind, and oppressed, it would be better, there would be more development. If only one is competent, it's not enough. This is my opinion. Now in our Nepal if there were more women who were educated like you, how would our Nepal be today? If I had opportunities to be educated, to stand on my own feet, it would have been something. When someone asks a woman to stand for [CFUG] chair, the heart says yes, but she has to take care of the children, goats, cattle, in-laws, her whole world, and eventually despite wanting to, the poor woman doesn't have time to take on those opportunities. *Didibahinis* [sisters; generic term for female peers] often tell me they would participate, but someone's husband might say no, another's mother-in-law . . . so even if they have opportunities, they cannot participate; not that they are not capable, but because they are unable to seize the opportunity.

(Bahun/Chhetri female, Chaturmukhi CFUG)

Nepal has come a long way since the days of *Sati* – when a Hindu widow would be expected to self-immolate in the husband's funeral pyre. Spivak (1988) has alerted us to the risk of White saviorism when there is no attempt made to understand the agency of the woman even in those circumstances. Women find ways to exert their agency in the most debilitating of circumstances and within their cultural constraints. The women I have known have – whether they were educated within Western institutions of learning or not – exerted autonomy even in limited ways within their spheres of control. Sometimes one woman's autonomy contributes to another's marginalization, whether in the same family or across the caste/Indigeneity divide. Women's oppression is often an extension of their own or other womens' agency within the patriarchal structures they have internalized. Beyond women's individual agency, there have been cultural strategies to protect women from *Sati*, as in the instance of the *Ihi* tradition observed by the Newar – where a girl is married to the *Bel* (wood apple) tree – a tradition that was practiced so that she could not be widowed and therefore subjected to *Sati*. The dowry tradition and all the elaborate obligations towards married daughters could be seen as a social structure to ensure daughters inherited parental resources even as they were excluded from inheritance of property.

As Nepal has become modernized, Eurocentric ideas of feminism have been introduced to raise women's visibility, participation, and mobility in the public sphere. Jayawardena (2016) cautions that such feminist interventions have tended to enhance women's freedom from traditional constraints only to render their labor exploitable within structures of capitalism, while stopping short of challenging structures of patriarchy. And hooks (2015, 86) has poignantly argued that "women cannot gain much power on the terms set by the existing social structure without undermining the struggle to end sexist oppression". There is a need to challenge patriarchal domination in Nepal in ways that do not presume oppressive non-Western traditional mores that appear oppressive to the Western gaze, as there may be irreversible losses from pursuing Eurocentric ideas of equality at the cost of erosion of Indigenous traditions built over thousands of years. But as Sen (1999) has argued, tradition ought not to be romanticized by outsiders, and those who carry the burden (and/or privilege) of maintaining tradition should be the ones to decide what is worth keeping and what to eschew. Well-intentioned external interventions on gender and other matters, although seemingly apolitical and simplistic, can be appropriated by women to suit their contextualized and particular circumstances, so the agency of the subaltern must not be underestimated (Shah 2018).

Feminist discourse in Nepal today includes but is not limited to issues including laws pertaining to property inheritance, citizenship rights, LGBT rights, *chhaupadi* (the spatial confinement of women during menstruation), calling out misogynistic and sexist representations in media, and the challenges of gender mainstreaming in development interventions. While supported by feminist discourse and gender equality imperatives from the West, much of this discourse has been organic, with many prominent women activists and political leaders leading the charge, even as it continues to be an uphill battle for them, given the entrenched patriarchal structures Nepal has inherited from a hegemony of Hindu Brahmanical precepts (Birkenholtz 2018). Despite progress in reforming laws to reflect greater gender equality – as in the provisions to secure 50 percent leadership positions in CFUG for women – Nepal still has a long way to go before women can pursue their potential and dreams of independence to the fullest. The gap between policy and practice has proved to be a great hurdle, and reverberations of this generic pattern are felt within CF and REDD+ bureaucracies. Therefore even if 50 percent of committee positions in CFUGs are occupied by women, power is not shared equally. Gurung and Gurung (2018) note that REDD+ consultative processes and forestry institutions have not met the standards set by the international Convention on the Elimination of All Forms of Discrimination against Women, even if it has been ratified by

Nepal. Women's empowerment has been largely understood in the limited sense of 'inclusion' as opposed to confronting the power dynamic, which is why REDD+ gender interventions appear to have been limited to 'counting bodies' while women's needs and agency haven't counted as much, in the everyday life of CFUGs. As Acha (2019) argues, beyond the mainstream narrative of women's empowerment seen in terms of victimhood and inclusion, there is a need to reveal the ways in which a patriarchal-capitalist structure keeps women in subordinate positions while exploiting their labor to aid in capitalist accumulation. The case of Chelibeti CFUG appears to be an exception to this pattern, and its existence holds promise for more substantive empowerment of women in the context of CF. But it appears that even this most empowered group of women in Chitwan's CFUG landscape has met a glass ceiling: "We are the ones who can't win. Finally we have risen up to these positions of authority, now these privileges are seen as too much and they limit our benefits" (Interview, Adibasi/Janajati female, Chelibeti CFUG chair, Chitwan, July 2019).

As in other spheres of life in Nepal, forest stewardship is gendered work, with women often performing most of the labor-intensive unpaid work, and men dominating decision-making work. This is true in the contexts of both CF and the household, as a result of which women's autonomy is significantly curtailed. These gender disparities are often chalked down to disparities in land tenure among men and women (Gurung and Gurung 2018), but cultural factors cannot be ignored either. Nepali women's agency, voice, and mobility tend to be controlled by their fathers, brothers, and/or husbands, and this often limits their exposure to professional development opportunities even when these are available. Wagle, Pillay, and Wright (2020) note that even as progress has been made in reforming formal institutions, their ineffectiveness in practice owes to informal institutional norms that hold women back. Women don't have genuinely equal power in forestry because "the 'inclusion of women' . . . has been largely rhetorical"; this state of affairs is not surprising when seen from the perspective of forestry insiders who divulge that "in Nepal, entry of women into the forestry profession began as an attempt to regulate women's behaviour in community forestry" (ibid, 259). As a result, even as a few – typically Bahun/Chhetri – women have overcome barriers to attain leadership positions, they are exceptions rather than the norm, having done so despite the structural constraints, not because of enabling structures. The structural constraints have proved to be largely insurmountable for women from Dalit/Adibasi/Janajati groups. These patterns of patriarchal domination are not unique to community forestry or to Nepal and have been discussed in the context of UNFCCC governance (Olson 2014), environmental and EJ work in the United States (Buckingham and Kulcur 2009), in US society at large, as well as in the lives of Black women (hooks 2015). Such patriarchal domination is reflected in epistemology as well – in the tendency to make universalizing claims about the causes of inequality (Chiro 2008; hooks 2015; Wright 2006); in the scientific method that has been used to fuel capitalist extractivism (Merchant 1983); and in the invisibilization of women's expertise (Acha 2019). While patriarchal mores abound everywhere, and they are possibly most oppressive in relatively more advanced capitalist societies (Peet and Hartwick 2015), the gender sensitization that arrives as part of environment and development interventions in Nepal – such as those in REDD+ social safeguard directives – is welcomed as an opportunity by/for Nepali women to exercise political agency within their communities, as well as to push the state to enact more gender-sensitive governance by enhancing opportunities and spaces for women, by changing the patriarchal cultures and structures of forest bureaucracies, by pushing for "a collective feminist voice" that includes the production of "gender-sensitive males" as well as top-down sanctions for pervasive entrenched male dominance often exhibited at community and district levels of governance (Gurung and Gurung 2018; Shah 2018; Wagle, Pillay, and Wright 2020, 260). But addressing gender independently from other forms of oppression is inadequate given how pervasive anti-Dalit/Adibasi/Janajati sentiment is among the most subaltern Bahun/Chhetri women as well as the culture of silence on gender within Indigenous leadership and activism in Nepal, even as patriarchal domination is less extreme in Indigenous communities than in Hindu communities (Rai 2020).

> For Indigenous women, the purpose and meaning of feminism is different from how it is viewed by the mainstream because they are acutely aware about how they have been robbed of their original customs, tradition, language, culture, Indigenous knowledge, skills and ancestral land due to the patriarchy fostered by Hindu norms and values. Therefore, the patriarchy and male dominance are not the only problems they have to deal with.
>
> The abolition of patriarchy and equal rights are certainly important, but they are not main agendas for them. The restoration of their unique identity and customary laws is the main agenda for Indigenous women. They want their unrestrained access to and control over their forest, water resource, land and ancestral habitat. Their struggle is for the recognition of their ethnic and gender identity, and their quest for the identity of Indigenous women is shaped more by the belief in Indigenousness than feminism. Feminism alone, as propounded by the mainstream, is not sufficient to address Indigenous women's gender issues because it fails short of helping their quest for Indigenous identity. Therefore, Indigenous women's movement should come up with a fusion of feminism and the belief in Indigenousness.
>
> (ibid)

A structural analysis of caste and Indigeneity in Nepal

I have noticed (and experienced) – as have others – the entrenched culture of patriarchal and caste domination and control in REDD+ programming as well as CF, and more generally in Nepal. The most visible manifestation of this is in the occupation of the most authoritative positions at multiple scales of bureaucracy including state (ministry to DFO), CFUG executive committee, as well as

FECOFUN, by Bahun/Chhetri males. REDD+ has prioritized gender equity and FPIC particularly for Indigenous groups as part of their commitment to Cancun Safeguards. Beyond this, Nepal has signed international treaties such as UNDRIP, the ILO Convention 169, and the Paris Agreement – all of which reiterate the recognition and rights of, and protections for, Indigenous Peoples – consequently assuming the status of domestic law (Gurung and Gurung 2018). However, the chasm between law and implementation has proven to be quite deep. As with gender, real change in hierarchical power dynamics cannot be engendered for disparities in Nepal by caste, ethnicity, or Indigeneity simply by formulating laws or by allotting resources based on the demographics of a community. While these are important measures – and should be protected and enhanced in terms of preserving seats for historically marginalized caste and Indigenous groups until critical mass is attained – understanding the structures of inequality is equally if not more important.

Structure here takes the form of formal and informal socio-cultural institutions and identity configurations that have been shaped by Nepal's history as a Hindu Kingdom and that are in the process of shifting and becoming reconfigured as it seeks to reinvent itself as a secular Federal Republic. Similar to the structure of racism in the United States, the caste structure, broadly speaking, shapes access to and power over resources, and its analysis is critical to any political ecology exploration in the South Asian context. Just as race is not a given category – nor a biological fact – but rather has been forged in a historically contingent process of "racial formation" (Omi and Winant 2015), so is caste not fixed, static, or a given. Caste is not entirely and clearly distinct/distinguishable from Indigeneity either (Damodaran 2006; Morrison 2006). The terms 'Dalit', 'Adibasi', and 'Janajati' are not natural categories with obvious meanings, nor is it possible to place everyone in Nepal into one of these categories neatly. These labels emerged in the context of a newer ethnic politics as a counterhegemonic struggle against Bahun/Chhetri dominance and in the process reflect an attempt to represent hybridized and complex identity formations into neat labels (Gellner 2019), much like the people of the region could not be neatly divided into distinct sovereign territories as cartographic maps would have us believe (Michael 2020). The formation of Adibasi/Janajati identity has been as complicated and contingent as that of caste (Gellner 2019).

In Nepal, caste formation has taken place both alongside *and* replacing Indigenous identity. Nepal's caste structure and its invisibilization of Indigenous Nationalities owe to an active process of Hinduization and Sanskritization that started at least in the first millennium A.D. and continued in waves in subsequent centuries but became congealed as a hegemonic structure with the formation of the nation state in 1768–9 (Bista 1991; Whelpton 2005). Nepal's Indigenous people are mainly categorized into two broad groups historically, the Khas and Kirat, that were known to have migrated and mingled since at least the early part of the first millennium A.D. Over time they formed a wide range of linguistic, ethnic, and cultural configurations with particular customs and traditions, many of which have persisted over time, often co-evolving with new cultural influences that arrived from the North (Tibeto-Burman) and the South (Indian). Caste-based hierarchies started to seep into the local culture with the migration of nobilities, along with their Hindu priests flee-ing Mughal invasions in their own territories to the South. They came as refugees and assimilated into the local traditions while successfully planting their Hindu logics of caste hierarchy, in which they, of course, belonged in the topmost echelons. Not all Indigenous groups in Nepal succumbed to Hindu caste hegemonies, but those who did would later be referred to as Hindu, their Indigenous traditions would co-evolve with Hindu traditions, and they would be folded into a caste hierarchy (Bista 1991). For those Indigenous groups that welcomed Hinduization and the caste structure, their public identity would be subsumed within the Hindu caste structure, even as their Indigenous traditions continued to flourish within the sphere of their private lives, suggesting that caste hegemony was never complete, and subjugated groups always found ways to exert their agency through subtle everyday acts of disobedience even as they might adopt the tools of the oppressors for survivance (ibid; Vizenor 1999). These incursions occurred in various places in current-day Nepal in particular ways, and hence there are many local variations of the complex caste structure.

It was with the unification project of the Nepali nation-state that the Hinduization project would near its apogee. Prithvi Narayan Shah, who main-tained connections with a royal lineage that had fled religious persecution in Rajasthan, was – like other similar entrants to current-day Nepal – intent on protecting Hinduism from a certain death, and in the process Shah, who was also motivated by the desire to thwart an impending British invasion, would create the "*only* and *last* truly Hindu country" (Gellner 2016, 13). Shah would take over the Newar Kingdoms of the Kathmandu Valley and suppress Newar language, customs, and literature, while appropriating their land, art, and cul-ture; Hinduism did not have to be introduced here per se, because it had already been planted by prior intrusions in more subtle and insidious ways. But what happened for the first time after the 'unification' was the codification of Hindu scripture and ideology in the new nation's laws, in the form of the *Muluki Ain* of 1854. This law asserted a one-language, one-identity, one-national-dress code of conduct for all of conquered territory now known as Nepal, hence imposing the caste hierarchy far and wide. Rules about enslaveability and degree of purity and untouchability based on the consumption of alcohol would be instituted. Those groups that resisted categorization or deemed casteless would be labeled Janajati. It was only since the 1990 Constitution of Nepal that caste-based discrimination was prohibited by law (Ellingson 1991), but by then the caste hierarchies and associated power geometries had become entrenched in the country.

Akin to the formation of Whiteness and White supremacy in the United States (Omi and Winant 2015), a supremacy of Brahmans – referred to as Bahunism in Nepal – was well established (Bista 1991). Although caste is typi-cally understood as a system of inheritance – and it is in large part how caste operates today – and due to its association with occupation, I would argue that caste and class categories are often co-created, although caste has elements beyond economic wealth and functions as a marker of prestige or social capital as well. There was also an active – and seemingly ad hoc – process of caste formation where Indigenous and immigrant groups would be folded into specific caste hierarchies by the ruling elite (often Bahun/Chhetri) based in part on their degree of assimilation and perceived professional contribution to the royal courts. It has also been documented that some groups adopted surnames that were perceived to be more prestigious and thus participated in their own caste formation. This happened at least in the case of Jung Bahadur Rana, originally of Magar ethnicity – that is, of Khas Indigeneity, comprising of a sizable proportion of Gorkha soldiers in Prithvi Narayan Shah's battalions that defeated the British as well as the Newar Kingdoms – who reinvented himself as a Chhetri Rana, associated with the Kshatriya or 'warrior' caste of Hindu nobility, and would go on to form the Rana regime that reigned for a century (ibid; Whelpton 2005).

Together, the Bahun and Chhetri castes, due to their close affiliation with the Monarchy, amassed much of the country's wealth through outright land seizure, *jagir*, and nationalization of forests, and today continue to occupy positions of power and authority over resources in the country through a process not too dissimilar from institutionalized racism in the United States (Ojha and Timsina 2008; Omi and Winant 2015; Regmi 1967). As a result, a clear majority in positions of power in government are Bahun/Chhetri, and the observed dominance in CF and REDD+ is simply a symptom of this wider malady that Bista (1991) terms alternately Bahunism and fatalism. Bista (1991) suggested two ways in which this malady has persisted in Nepali society: through a culture of *afno manchhe*, similar to nepotism but could extend to one's clan, caste, or otherwise favored group; and through an internalized culture of paternalism, wherein those wrongfully placed in the lower rungs of the caste hierarchy looked to higher-ups as a 'father figure', relinquishing their well-being to the mercy of their socially appointed patrons. He argued that the hegemony of the Bahun/Chhetri caste and the attempted erasure and suppression of the potentials of Nepal's Indigenous peoples is to blame for Nepal's 'underdevelopment', and that the Indigenous people of Nepal held the key to the country's future prosperity. The Maoist movement that originated in the 1990s was in large part a reaction to the perceived injustice of Nepal's caste hierarchy. In the ensuing decades, assertion of ethnic identity has become a growing political agenda, with the emergence of 'ethnic' categories that reflect historical marginalization: Dalit, those oppressed in the caste structure, and Janajati, referring to ethnic Nationalities (Gellner 2016).

The discourse of Indigeneity is relatively new in Nepal. Following the people's movement, NEFIN was created in 1991 as the Nepal Federation of Nationalities. Influenced by the International Decade of the World's Indigenous Peoples (1995–2004) discourse, the term 'Adibasi' – corresponding with Indigenous – was added to the Nepali version of the name, which then became the Nepal Federation of Indigenous Nationalities in 2003 (Magar 2020). During the transition to a Federal Republic, demands for ethnic representation in governance have grown, with 45 percent of civil service positions now reserved for marginalized groups: 33 percent for women, 27 percent for Adibasi/Janajati (with distinctions made between more and less disadvantaged), 9 percent for Dalits, and so on – a change that a newly constituted Khas/Arya (generally known as Bahun/Chhetri) category of people vehemently opposed (Gellner 2016, 2019). Even with these efforts, the struggle for true representation and to break away from Bahunism continues to be an uphill one – for instance, there is currently a reinvigorated effort to Sanskritize the school curriculum (Ghimire 2020) – the role of external forces, such as FPIC and the Cancun Safeguards, although not essential for the organic movements, can definitely bolster and lift them up, if addressed in a meaningful and genuinely representative spirit.

As REDD+ discourses enter Nepal's landscape of ethnic politics, they bring the formal recognition of the special role that Indigenous people and their traditional land use practices and institutions are known to play in contributing to ecological sustainability. They can thus in theory elevate the importance of protecting Indigenous knowledge, existence, and tenure rights; and in light of Nepal's ongoing ethnic struggles, the exclusive focus on the Indigenous becomes broader to encompass Dalit as well as other marginalized groups such as the Madhesis and Muslims (Sherpa, Rai, and Dawson 2018). At the national level, NEFIN has had a voice in Nepal's REDD+ discourse and it has supported it with caveats; DANAR, on the other hand, has been sidelined (Satyal et al. 2019). Although NEFIN has enjoyed a consultative role, the degree to which NEFIN as an organization is representative of the Indigenous people in the Terai is questionable, with a considerable gap in access to information and resources between the center and the periphery observed within this group. Meanwhile, ongoing marginalization, disenfranchisement, and invalidation of the claims of Dalit/Adibasi/Janajati groups continue among the ranks of CFUGs. Following are glimpses of conversation from the August 2019 forum.

> Janajati, Adibasi, Dalit are the ones who live near the forest. How can you ensure that their rights and their access to benefits can be guaranteed? I don't think this carbon trade is good. The agreement has been finalized. When were these people consulted? Where are the specific points about their rights? There is no clarity as to how this will be achieved. Even we in leadership positions [in CF] are confused. Carbon trading will be like other international programs. The government enters into agreements. The government benefits. Decisions are made at the top level, but the forests are being conserved at the ground level. Will the benefits reach the ground level or not? There is a lot of noise around carbon trading. But the proceeds won't go to the bottom levels.
>
> (Dalit female from Kapilvastu)

> Janajati, Dalit, these are divisive issues. We are all Nepali. We should move beyond Janajati/Dalit issues. We shouldn't get bent out of shape over issues of *jaat* [caste/ethnicity].
>
> (Bahun/Chhetri male, Gandaki FECOFUN)

[He goes on to argue that the quota system of reserved seats is unfair. In his youth he had been quite demoralized when he was not selected for a *lok sewa* government position, leaving him to doubt his abilities. Another person responds . . .]

There is no equality for Janajati people. There is a lot of disparity. They have been left behind. Look at all 77 districts. How much representation is there for Janajati? Are you saying they are all equal? No, there is much inequality. You have to accept, Sir. We don't want anyone's rights to be

Figure 6.3 Non-timber forest products harvested by CFUG members in Gorkha, August 2017

harmed but we need to have equality. The government is trying to remedy, it is being debated, but we need to understand this. Us Janajati are only seeking our rights. Not seeking to snatch yours. That is all . . . Even when qualified, the positions [*lok sewa*] go to *afno manchhe*. That's the bitter truth. Chhetri/Bahuns occupy all the positions. Because of inequality we are seeking justice.

(Adibasi/Janajati male from Banke)

Unpacking 'sustainable forest management': ecology and economy

Ecological concerns drove the first wave of radical reforms in forestry, leading to a (partial) taking back of the commons. It remains to be seen if the recent wave of reforms in community forestry will undo those gains and lead us (back) to a state of privatized forests, essentially reduced to 'carbon stock'. What is clear is that the most recent iterations of reform (reversals) are couched in the language of 'sustainable forest management' (SFM) mirroring the language of the Paris Agreement in reference to REDD+. Nepal has ushered in a whole host of Acts, Policies, and Laws that have at their center an endorsement of SFM as the new aim around which much forest governance will focus, seemingly to prepare for alignment with carbon financing modalities, keen to produce " 'deforestation free' commodities that could provide a boost to the domestic economy and support sustainable forest management through commodity supply chains" (MoFE 2019, 22). It is therefore timely to revisit the question of what 'sustainable' means – as was broached in Chapter Four in the context of biogas projects for the Clean Development Mechanism – and to revisit the notion of 'sustainable livelihoods' as "intimately related to rights of communal ownership" (Sheppard et al. 2009, 150). A sustainable livelihoods approach to sustainability is compatible with strengthening the resilience of forest communities in the face of climate perturbances: "A livelihood is sustainable which can cope with and recover from stresses and shocks, maintain or enhance its capabilities and assets, and provide sustainable livelihood opportunities for the next generation" (Chambers and Conway 1992, 36–37). In a context of colonial usurpation of resources – which I argue the creation of the Nepali nation-state represents – I suggest the pursuit of environmental justice should focus not only on distributive and participatory justice but also on structural changes that strengthen self-determination, self-governance, and autonomy in the pursuit of sustainable livelihoods. This is essential in order to enhance the capabilities of disenfranchised groups and to undo the damage done particularly to Indigenous groups by colonization of their resources and territories. If colonialism enclosed the commons and neocolonialism/neoliberalism threatens to privatize it, then decolonization should mean returning the commons to their rightful users. This would be compatible with the Cancun Safeguards and the UNDRIP's commitment to "guarantee Indigenous Peoples' rights to land, territories, resources, ancestral domain, and their right to self-determination and cultural integrity" (Sherpa, Rai, and Dawson 2018, 75).

However, as a 'floating signifier' the term 'sustainability' tends to be used in more reductionist ways and has indeed been equated with increasing productivity for the nation's prosperity in the new MoFE legislative discourse. In this sense, the meaning of sustainability is closer to the notion and practice of 'sustained-yield forestry' that can be traced to the 1713 work *Sylvicultura Oeconomica* by German forester Hans Carl von Carlowitz, following which forest management planning and experimental forestry were introduced to Europe in the 19th century, and not without conflict: "However strongly a science-based reform programme tries to disentangle itself from the politics of nature, such programmes, as well as the terminology by which they are fed into and received by public discourse, are subject to historical change and power struggles" (Holzl 2010, 431). In my conversations with CFUG members and state officials, I recognized a matter-of-fact sense among state officials about the superiority of a 'sustained-yield' approach towards both timber and carbon production, labeled 'scientific forestry' (SF), encompassing a view of forests as farms; CFUG members were divided, and not along neat lines of gender, caste, and ethnicity/Indigeneity. In the local context SF was generally seen as a departure from subsistence use-oriented traditional forest management practices, which felt threatening to some and not to others, particularly those who were interested in generating more revenue from community forests. This change was no doubt driven by state agencies; has been adopted by a small percentage of CFUGs; and, given the variety of responses, appears to have been practiced differently in different places.

> We have been conserving forests in Nepal since ancient times, following traditional practices, not [SF]. We should stick to traditional methods. The Ministry of Forests has been pushing [SF]. [CF]s have been against it. Maybe this is related to carbon. Earlier we used to make do with only cutting dry and fallen trees. Now they want us to work in 80-year rotations, cutting all but the *mau* [mother] tree. Vultures might go extinct . . .
>
> (Bahun/Chhetri male, Jaldevi CFUG).

> Everyone has their own opinion about [SF]. We are divided. From my point of view, I don't like [SF]. I like natural forests. Where there are no forests perhaps SF is good, but I don't like felling natural forests for the sake of [SF]. The soil is different in different areas of forest. If you cut down the *Sal* forest and plant other, expensive *jaatko* [varieties], we don't like it . . . With [SF] not only the old, fallen and dried trees are cut but the entire landscape in the name of forest management. Then they plant something else. Naturally there were different varieties, they grow on their own, that is better. We would remove the old, fallen and dry trees for our use. That was working just fine.
>
> (Adibasi/Janajati female, Chelbeti CFUG)

> I can see it both ways. Where there is a lot of growth, managing scientifically is good. The old trees should be removed. Then optimal growth

conditions can be achieved. We should keep intact ancient forests but where growth is new, managing for optimum conditions of growth is good. Keeping a few and removing the old ones is good.

(Adibasi/Janajati female, Amritdhara CFUG)

I was against [SF]. But when I saw that they weren't cutting down healthy trees, that it was just the old, shabby, and hollowed out trees, I was convinced. We started [SF] last year. The forest is divided into eight plots. Each plot is harvested over 10 years to meet community needs while the other plots are left to regenerate.

(Adibasi/Janajati male, Baghdevi CFUG).

In an interview with the same gentleman the year before, he had shared that his CFUG had after a long period of recalcitrance tried SF and in the process had engaged in the planting of the *chiuri* tree, an important species for the Chepang, who value it so deeply for its many livelihood benefits – insecticidal, medicinal, edible fruit, *ghiu* (ghee), bee-keeping, furniture, and so on – that in the old days the tree would be gifted to newly wed daughters as dowry. He appreciated the systematized techniques and deliberate planning of SF, but a problem was the need to hire technical experts at the district level when locals were themselves "no less than doctors for forests in terms of expertise" (Interview, August 2018, Chitwan).

Forest users are therefore not necessarily against the idea of productive forests and prosperity for their communities. While critical scholars often critique the oxymoronic nature of the phrase 'sustainable development' for its easy association with economic growth – an ugly proposition for many ecologically minded thinkers – for some CFUG members, the notion that stewardship of forests may lead to economic gains was an attractive one. Community forests and REDD+ in this context are seen as a means to upward mobility. While ecologically minded thinkers tend to equate Indigenous people and forest dwellers with a subsistence-oriented lifestyle deemed 'traditional' and 'sustainable' – perhaps an imaginary of green Orientalism (Lohmann 1993) for both the West and the Westernized urban center of Kathmandu – Pokharel et al. (2008) have argued that CFUGs have found this presumption problematic, and they have resisted the early state discourse that community forestry was meant only for subsistence. My conversations with forest users also reveal an affinity towards a money economy and aspirations for modernist development and technical capacity-building, as well as an eagerness to derive financial gains from the sale of forest products, including but not limited to carbon, alongside a wistful resignation to the inability to profit from the sale of timber. Results-based financing is therefore seen as a welcome opportunity to diversify the local economy. The REDD+ seed money had been used by CFUG committees as a sort of rotating fund bank offering zero interest loans, thus serving as a sustainable source of money for investment in income-generating businesses. Among critical scholars of development there has been a desire to dissociate development from economic growth in critical development thinking and to think of development as the enhancement of basic freedoms and capabilities (Lele 1991; Sen 1999). But as Sen (1999) was careful to point out, the call to go beyond economic well-being did not mean economic growth and well-being were not desirable or important for the fulfillment of capabilities, and he cautioned against a tendency to romanticize poverty and underdevelopment. It appears that from the point of view of forest users, there is not a contradiction to seeing forests as commons and as a possible source of commodities, as long as the financial gains benefit the community and are not siphoned away to distant centers of power, whether the state or foreign entities. So the questions of whose forests, whose wealth, and who decides what is sustainable and for whom appear to be the central ones as opposed to whether scientific forestry or economic progress is intrinsically good or bad.

Beyond who controls the resource base, it matters how productivity and economic well-being are understood, counted, and measured. European colonialism and subsequent neocolonialisms have not only participated in and enabled resource usurpation but have created a hegemony of extremely narrow interpretations of resource productivity and economic progress that have been internalized by resource-based communities through governmentality (Agrawal 2005). These reductionist epistemological approaches for translating ecological and human capability into highly particular yet universalized notions of progress have been diffused around the world through colonialism and subsequent economic globalization that has incorporated far reaches of the globe's people and resources into a circuit of international trade. As a result, not only have resource-dependent communities in pre-modern societies often become disenfranchised from their resource base, their knowledge systems and traditional expertise have been devalued and invalidated. As such SF and 'neoclassical economics' have come to be associated with colonialism (Robbins 2004; Sheppard et al. 2009). SF was essentially colonial science as practiced in British India (Agrawal 2005; Rangan 2000). What Nepali scholars and practitioners of community forestry term techno-bureaucracy of the state's forest agencies is perhaps a continuation of the same. The technical requirements of quantification for carbon legibility, particularly for MRV, appear to be compatible with this techno-bureaucratic approach to forestry (Osborne 2015).

The challenge for REDD+ implementation in Nepal will be to balance these imperatives against the Cancun Safeguards that require respect for and preservation of traditional knowledge and practices of affected Indigenous groups. As REDD+ unfolds in the Nepali Terai, it will be interesting to see if traditional knowledge holders may be able to rework SF as colonial science in a Bhabhaian sense to advance their goals of autonomy and access over the resources their livelihoods depend on. In this sense, what 'science' and SF mean may need to be opened up for debate, for it is not as if traditional societies have not practiced rigorous studies and measurements of the natural world; rather, their knowledge systems have been systematically dismantled and devalued to a large extent under colonialism. Equally important is to refrain from presuming that Indigenous people are necessarily predisposed to reject Western science. Rather, it is critical to consider who controls scientific study and for what purpose. Baghdevi CFUG members who had embraced SF but had begrudgingly worked with external 'experts' revealed, for instance, that it is not the technique of selective harvesting they object to, but rather that their choices in selection of *mau* trees and others needed to take precedence. They appreciated the technique but wanted to be the ones to control it (Interview, Adibasi/Janajati male, Baghdevi CFUG, August 2018).

Opposition to SF among CFUG members appears to be in response to its perceived focus on high-value timber and carbon stock – whereas to local forest users the productivity of a wide variety of plant species is desirable – but also because of exclusive control of this 'colonial' tool by the state. The myth of the *Chipko* movement comes to mind – Rangan (2000) effectively argued that the famed women of the *Chipko* movement in the Garhwal Himalaya were hugging trees not to *protect* them from the loggers but because *they* wanted control over forest resources. Similarly, even as SF is debated within CFUGs, there is growing interest in the potential of CF to generate more revenue through adoption of SF, with key concerns focused on possible loss of autonomy over decision-making and revenues. Over three summers of research, I have met numerous CFUG members, mostly Bahun/Chhetri men, but also many women of various ethnicities, who were eager to learn the ropes of carbon measurement and enhancing their technical capabilities so that they could use those tools and resources to keep forest-generated wealth – that would otherwise be siphoned off to outside experts – in their communities. These aspirations also suggest that rural forest users are aware of the deficit lens through which forestry bureaucrats from Kathmandu view them, and seek to overcome these perceived deficiencies by securing access to new knowledge pertaining to REDD+ and climate change. While these desires concerning modern techniques and technologies should be met adequately by the state – so that deficit thinking cannot be used as an excuse to limit genuinely deliberative processes – it is also necessary for the state to do its due diligence in complying with its commitments to the Cancun Safeguards, and the related obligations to uphold and recognize the value and validity of Indigenous ways of knowing and managing the natural world. Decolonization here entails sharing power through devolution of governance, as well as loosening the hegemony of limited and limiting notions of sustainability, productivity, and progress, while avoiding the tendency to freeze the Indigenous and the rural resource user into a pre-modern subsistence imaginary. The most ethical approach appears to be to enhance the agency and self-determination of resource users themselves in navigating the choices between tradition and modernity as they seek to make a livelihood while being enveloped by globalizing forces and discourses.

It has been five years since [SF] was introduced . . . [SF] policy is made by technical experts. They have to be paid, they don't work for free. They want [CF] to include [SF] in their working plans but *we* have to hire experts. The ordinary person does the difficult work but we have to employ experts . . . have to pay 18–20 lakhs once we transition to [SF]. The forest yield has to be analysed, how much wood has to be cut. Do you have to ask me if you want to harvest and sell your wheat? To sell your buffalo? No. [CF] has conserved forests and we get to use for our daily needs, but to sell outside we have to seek permission. Do you seek my permission to take your bicycle out? It's your bicycle. But here you have to get permission to use your own bicycle. Okay, you have donated your cow to someone. Does the recipient have to get permission to milk it? To drink the milk? You are rearing it. So when you adopt [SF], you cut trees, you auction it. But we only get a percentage. You have to give a cut to the higher ups. . . . Now whether it is [SF] or REDD+, the rate is so low. It's only in Nepal where the rate can be so low. Where else would people protect their forests with so much love? And now when that money comes, it goes to the state's treasury . . . 3–4 years ago I met a FECOFUN official. We discussed how when the carbon money comes and a pipeline is created for money to flow to [CF], only a small part, if at all, will arrive, after much leakages (woman: right, only leftovers) and they say in the government's action plan, we are rich: *Hariyo Ban Nepalko Dhan*. . . . The government hasn't invested in [CF] and how can they claim the benefits? How can they tax us? 40 to 90% . . . If REDD+ brings some money, the government will do the same. They won't give the money to us straight. But if they don't, the people will create trouble. . . . If you are living in your house but are not allowed to do anything freely, you wouldn't want to live there. You would just want to leave. You have to have something.

(Bahun/Chhetri male, Amrit Dharapani CFUG, July 2019)

Concluding thoughts

Nepal's community forestry offers a case of a commons that is being woven into the fabric of the atmospheric commons. In this forest commons, resource users have control over their means of production, albeit limited, with the state retaining ultimate ownership and rule-making authority. CFUGs make community-level rules about jurisdiction, resource use, sanctions, budgets, and everyday governance. This arrangement meets most of their subsistence needs and increasingly supports community development through infrastructure projects and economic opportunity. This level of community autonomy and self-governance has been won through struggle, and state-CFUG power dynamics continue to evolve, navigated by civil society groups and waves of international intervention. The first wave of intervention strengthened community autonomy at a structural level; the new wave of REDD+ related intervention brings conflicting prospects for the future of community forestry. On the one hand, there are explicit equity priorities that so far have been pursued in well-meaning but tokenizing ways. There is a significant opportunity to strengthen the struggle for Indigenous self-determination over forests at the community and national levels on the grounds of the Cancun Safeguards, if taken seriously. If these commitments are treated as an inconvenient obligation for the state to meet by doing the absolute minimum level of token consultations, there is a risk of perpetuating the historical dispossession of the agency, autonomy, and knowledge of Indigenous and other marginalized forest users, damaging the legitimacy of REDD+. Such dispossession is concerning in and of itself, but when forest carbon becomes brought into the fold of global climate mitigation agendas and obligations, the import of dispossession is multiplied in the global context of profit accumulation. Until 1951 Nepal was an isolated nation-state – kept deliberately so by the Shah-Rana regime to keep British colonial powers at bay. A mere seven decades later, forest-dwelling rural communities are now being ushered into a world of global carbon trade whether or not they gave their permission, and whether or not they fully understand the implications and risks of entering this global market – for their communities and for the planet.

Yet, examining the commodification of the atmosphere and that of forests simply as a project of global capitalism is inadequate. Political ecology has long had a staunch commitment to Marxist visions of justice and equality. Many scholars have pointed to a need to move beyond a strict economic determinism – which is yet another form of dominance – to incorporate more substantive analyses of gender, race, and coloniality (Acha 2019; Bryant 2001; Grosfoguel 2011; Laclau 1990; Omi and Winant 2015). As the Nepal case shows, we cannot understand the relationship between power and resources in the South Asian context and the ways in which resource usurpation and control have taken place in non-Western societies since precapitalist times, without moving beyond a US-centric understanding of race and racialization. As Gellner (2016, 31) notes, applying generic ideas of caste is not adequate either:

> It is a common mistake for outsiders who know a little about Nepal to imagine that ethnic affiliation or caste determines everything about Nepali society, that once you know a Nepali's caste, you know everything about them. It is an equally common error (among economists, or migration experts, for example) to think (or assert, or just assume without even realising it) that ethnicity means nothing – all that matters is poverty and economics.

I would add to this an indiscriminate application of a lens of Indigeneity from settler colonial contexts such as the United States to places like Nepal. Centering European understandings and theories of the world, including notions of economic equality and Indigenous struggle, in a much more diverse and expansive world, is yet another mode of dominance. This particular dominance occurs in a way that invisibilizes distant others from the point of view of the Euro-US center. Doing so precludes mutual learning that can happen for the common good.

When Third World countries are *only* seen as places of strife and where White people can *only* act as saviors – whether by giving development aid or by offering purist notions of radical change or justice, e.g. socialism – then the world loses, because there is so much the First World can *learn* from the Third World. In *The Dispossessed*, Le Guin (1974) imagines an ideal-type futuristic world Anarres, a society based on mutual aid, in which there is no corporal punishment and consequences are meted out via shame culture; while another, Terra, has characters described as yellow-brown, who join their hands as in a *Namaste* and who sit on the floor, who came from an overpopulated Earth that has already met its tragic neo-Malthusian demise. In Hahnel (2012), the ideal world is discussed in theoretical and hypothetical terms with no connections made to actual societies elsewhere that actually follow similar practices. In these two examples, I find a simultaneous erasure and appropriation of Third World cultures when First World thinkers pass off their visions as futuristic, hypothetical, and not borrowed from recent pasts and distant presents. The idea of the future as imagined rather than inspired by real communities striving for justice – albeit in imperfect ways – is made possible by the invisibilization of the latter. Such versions of Marxist political economy, I suggest, offer a new guise of White saviorism. It would behoove critical scholars to discard their US- and Euro-centric tendencies and treat Indigenous people, practices, and knowledge – in Euro-American settler colonial contexts as well as other contexts – as valuable and worthy of learning from – as opposed to be proselytized and patronized. Nepal's community forestry, I believe, is an example to learn from if not emulate. It shows how a structural change based on a vision and practice of the commons is possible at the national scale, even as it is not perfect, particularly now that it is enmeshed with an atmospheric commons that is being commodified, and even as the struggle continues and may never be over. The key battle of the contemporary moment may be between these two forces, commoning and commodification. What is important here is to recognize that community forestry represents one of "the plurality of social agents" (Laclau 1990, 130) and forms of "socialization of power" (Grosfoguel 2011, 5) that may, if scaled up, build the kind of counterhegemonic movement that may successfully supplant the current hegemony of state-centric and Euro-centric colonial capitalism.

Centering the desires and aspirations of forest users participating in the struggle helps us appreciate the complexities of some cherished truisms in critical thought about environment and development. Critiques of economic determinism in notions of well-being are warranted when the rhetoric of economic development and growth is appropriated by the state to concentrate wealth accumulation in the hands of the few. But economic progress and growth are valued by local resource users, even as from a Westernized environmental imaginary, forest dwellers are Indigenous people who have a special relationship of oneness with nature and should remain in that rarefied space. Unlike in contexts where rampant extractivism is the norm, the members of the forest commons have shown their ability to steward their resources while utilizing them as commodities for their economic prosperity. Following hooks (2015), who urged a view of feminism as a struggle against *patriarchal domination* as opposed to hatred of men, I suggest that a way to engage with questions about science, markets, development, and sustainability is to emphasize and problematize the domination of these social spheres by powerful countries and entities and to seek to understand the ways in which subaltern communities work with their prevailing constraints to challenge relations of domination, as opposed to seeing them as the problem. If colonial patterns of domination offer more power and privileges to dominant groups – such as Global North countries, state authorities, or dominant groups in society – in controlling resources, discourse, and policies, then it is the power dynamic that is important to analyse and unpack. The perspectives of those who labor to protect forests and biodiversity, and who bear the brunt of risks created by Global North privileges, must be taken into consideration in formulating solutions to our global crises. The people I have heard from have conveyed in no uncertain terms that they wish to be active agents and not passive recipients in contemporary discourse and action on climate change. They are among the last people on earth to be patronized about how to mitigate climate change or manage their forests. While the agency and empowerment of CFUGs in Nepal is to be applauded, what does appear problematic is the entrenched dominance of Bahun/Chhetri males throughout the forest bureaucracy, FECOFUN, as well as CFUGs. Here external interventions such as REDD+ may enable and bolster marginalized individuals and groups from household to national scales to seek equitable access to resources in an environment of entrenched inequality. To a certain extent this has happened for some marginalized groups and not others. The legitimacy of REDD+ rests on whether hitherto marginalized Dalit/Adibasi/Janajati/women are enabled by its interventions to be autonomous agents in the context of Nepal's forest commons.

References

Acha, Majandra R. 2019. "Climate Justice Must Be Anti-Patriarchal, or It Will Not Be Systemic." In *Climate Futures*, edited by Kum-Kum Bhavnani, John Foran, Priya A. Kurian, and Debashish Munshi. London: Zed Books.

Agrawal, Arun. 2005. *Environmentality*. Durham: Duke University Press.

Allen, Charles. 2015. *The Prisoner of Kathmandu*. New Delhi: Speaking Tiger.

Barrios, Roberto E. 2016. "Expert Knowledge and the Ethnography of Disaster Reconstruction." In *Contextualizing Disaster*, edited by Gregory V. Button and Mark Schuller. New York: Berghahn Books.

Birkenholtz, Jessica V. 2018. *Reciting the Goddess*. New York: Oxford University Press.

Bista, Dor B. 1991. *Fatalism and Development*. Hyderabad: Orient BlackSwan.

Blaikie, Piers. 1980. *The Political Economy of Soil Erosion in Developing Countries*. London: Longman.

Blaikie, Piers, and Harold Brookfield. 1987. *Land Degradation and Society*. London: Methuen & Co. Ltd.

Blaut, James M. 1993. *The Colonizer's Model of the World*. New York: The Guilford Press.

Bryant, Raymond. 2001. "Political Ecology: A Critical Agenda for Change?" In *Social Nature*, edited by Noel Castree. Oxford: Blackwell.

Buckingham, Susan, and Rakibe Kulcur. 2009. "Gendered Geographies of Environmental Justice." *Antipode* 41, no. 4: 659–83.

Bumpus, Adam, and Diana Liverman. 2010. "Carbon Colonialism? Offsets, Greenhouse Gas Reductions and Sustainable Development." In *Global Political Ecology*, edited by Richard Peet, Paul Robbins, and Michael Watts. London: Routledge.

Chambers, Robert, and Gordon R. Conway. 1992. *Sustainable Rural Livelihoods*. West Sussex: Institute of Development Studies, February.

Chiro, Giovanni Di. 2008. "Living Environmentalisms: Coalition Politics, Social Reproduction, and Environmental Justice." *Environmental Politics* 17, no. 2: 276–98.

Cole, Luke W., and Sheila R. Foster. 2000. *From the Ground Up*. New York: New York University Press.

Damodaran, Vinita. 2006. "Indigenous Forests." In *Ecological Nationalisms*, edited by Gunnel Cederlof and K. Sivaramakrishnan. Seattle: University of Washington Press.

Dhungana, Sindhu, Mohan Poudel, and Trishna S. Bhandari, eds. 2018. *REDD+ in Nepal*. Kathmandu: REDD Implementation Centre, Ministry of Forests and Environment, Government of Nepal.

Ellingson, Ter. 1991. "The Nepal Constitution of 1990: Preliminary Considerations." *Himalaya* 11, no. 1: 1–19.

Estes, Nick, and Roxanne Dunbar-Ortiz. 2020. "Examining the Wreckage." *Monthly Review*, July 1.

Federici, Silvia. 2004. *Caliban and the Witch*. New York: Autonomedia.

Foster, John B. 1999. *The Vulnerable Planet*. New York: Monthly Review Press.

Gellner, David N. 2016. "The Idea of Nepal." The Mahesh Chandra Regmi Lecture 2016, Social Science Baha, Kathmandu, December 11.

———. 2019. "Masters of Hybridity: How Activists Reconstructed Nepali Society." *Journal of the Royal Anthropological Institute* 25, no. 2: 265–84.

Ghimire, Binod. 2020. "Government Decision to Introduce Sanskrit in School Education Draws Controversy." *The Kathmandu Post*, May 17.

Gibson-Graham, J. K., Jenny Cameron, and Stephen Healy. 2013. *Take Back the Economy*. Minneapolis: University of Minnesota Press.

Griffin, Paul. 2017. "The Carbon Majors Database." July. https://b8f65cb373b1b7b-15feb-c70d8ead6ced550b4d987d7c03fcdd1d.ssl.cf3.rackcdn.com/cms/reports/documents/000/002/327/original/Carbon-Majors-Report-2017.pdf.

Grosfoguel, Ramon. 2011. "Decolonizing Post-Colonial Studies and Paradigms of Political-Economy." *Transmodernity: Journal of Peripheral Cultural Production of the Luso-Hispanic World* 1, no. 1: 1–38.

Gurung, Jeannette, and Dibya Gurung. 2018. "Gender and REDD+ in Nepal." In *REDD+ in Nepal*, edited by Sindhu Dhungana, Mohan Poudel, and Trishna S. Bhandari. REDD Implementation Centre, Ministry of Forests and Environment, Government of Nepal. Kathmandu: Jagadamba Press.

Hahnel, Robert. 2012. *Of the People, by the People*. New York: AK Press.

Hardin, Garrett. 1968. "The Tragedy of the Commons." *Science* 162: 1243–48.

———. 1994. "The Tragedy of the Unmanaged Commons." *Trends in Ecology & Evolution* 9, no. 5: 199.

Harvey, David. 2018. *Marx, Capital and the Madness of Economic Reason*. New York: Oxford University Press.

Holzl, Richard. 2010. "Historicizing Sustainability: German Scientific Forestry in the Eighteenth and Nineteenth Centuries." *Science as Culture* 19, no. 4: 431–60.

hooks, bell. 2015. *Feminist Theory*. New York: Routledge.

IPCC. 2014. "Climate Change 2014: Synthesis Report." In *Fifth Assessment Report of the Intergovernmental Panel on Climate Change*, edited by Core Writing Team, Rajendra K. Pachauri, and Leo A. Meyer. Geneva: IPCC.

Jayawardena, Kumari. 2016. *Feminism and Nationalism in the Third World*. New York: Verso.

Klein, Naomi. 2008. *The Shock Doctrine*. New York: Picador.

Laclau, Ernesto. 1990. *New Reflections on the Revolution of Our Time*. New York: Verso.

Le Guin, Ursula K. 1974. *The Dispossessed*. New York: HarperVoyager.

Lele, Sharachchandra M. 1991. "Sustainable Development: A Critical Review." *World Development* 19, no. 6: 607–21.

Levy, Robert I. 2019. *Mesocosm*, 2nd ed. New Delhi: Motilal Banarsidass.

Lohmann, Larry. 1993. "Green Orientalism." *Ecologist* 23, no. 6: 202–4.

Magar, Suresh A. 2020. "Nepal Federation of Indigenous Nationalities: Moments of Formation." *Nagarik Khabar*, August 29.

Manandar, Ugan, and Charlie Parker. 2018. "REDD+ Finance in Nepal." In *REDD+ in Nepal*, edited by Sindhu Dhungana, Mohan Poudel, and Trishna S. Bhandari. REDD Implementation Centre, MoFE, Government of Nepal. Kathmandu: Jagadamba Press.

Manandhar, Supriya. 2018. "Ambitious Pride Projects Threaten Khokana Heritage." *The Record*, January 19.

Mandal, Chandan. 2019. "Taking Back Forests from Its Community Users Is Curtailing Their Rights, Activists." *The Kathmandu Post*, May 27.

———. 2020. "Community Forest Users' Groups Decry Continuation of Heavy Taxes." *The Kathmandu Post*, June 4.

Merchant, Carolyn. 1983. *The Death of Nature*. New York: HarperCollins.

Michael, Bernardo. 2020. "From Frontier to Boundary." *Himal Southasian*, August 26.

Mishra, Chaitanya. 1980. "Review Article. Nepal in Crisis: Growth and Stagnation at the Periphery." *Contributions to Nepalese Studies* 8, no.1: 223–28.

MoFE. 2019. *Submission of Voluntary National Report to United Nations Forum on Forests Secretariat*. Kathmandu: MoFE, Government of Nepal, November 29.

Morrison, Kathleen D. 2006. "Environmental History, the Spice Trade, and the State in South India." In *Ecological Nationalisms*, edited by Gunnel Cederlof and K. Sivaramakrishnan. Seattle: University of Washington Press.

Ojha, Hemant R., and Netra P. Timsina. 2008. "Community-Based Forest Management Programmes in Nepal: Lessons and Policy Implications." In *Communities, Forests and Governance*, edited by Hemant R. Ojha et al. New Delhi: Adroit Publishers.

Ojha, Hemant R. et al., eds. 2008. *Communities, Forests and Governance*. New Delhi: Adroit Publishers.

Olson, Jessica. 2014. "Whose Voices Matter? Gender Inequality in the UNFCCC." *Agenda: Empowering Women for Gender Equity* 28, no. 3: 184–87.

Omi, Michael, and Howard Winant. 2015. *Racial Formation in the United States*, 3rd ed. New York: Routledge.

Online Khabar. 2020. "NEFIN Opposes Decision to Teach Sanskrit." *Online Khabar*, May 8.

Osborne, Tracey. 2015. "Tradeoffs in Carbon Commodification: A Political Ecology of Common Property Forest Governance." *GeoForum* 67: 64–77.

Ostrom, Elinor. 1999a. *Self-Governance and Forest Resources*. Occasional Paper No. 20. Bogor: Center for International Forestry Research.

————. 1999b. "Coping with Tragedies of the Commons." *Annual Review of Political Science* 2: 493–535.

————. 2014. "A Polycentric Approach for Coping with Climate Change." *Annals of Economics and Finance* 15, no. 1: 97–134.

Paudel, Dinesh. 2016. "Re-Inventing the Commons: Community Forestry as Accumulation Without Dispossession in Nepal." *The Journal of Peasant Studies* 43, no. 5: 989–1009.

Paudel, Ramesh K. 2020. "Chepang Organization Demands Inquiry into Kusum Khola Incident." *The Kathmandu Post*, July 24.

Peet, Richard, and Elaine Hartwick. 2015. *Theories of Development*, 3rd ed. New York: The Guilford Press.

Pokharel, Bharat K., Peter Branney, Mike Nurse, and Yam B. Malla. 2008. "Community Forestry: Conserving Forests, Sustaining Livelihoods and Strengthening Democracy." In *Communities, Forests and Governance*, edited by Hemant R. Ojha et al. New Delhi: Adroit Publishers.

Rai, Kailash. 2020. "Feminism Through the Lens of Nepal's Indigenous Women." *Indigenous Voice*, August 24.

Rai, Rajesh, Basant Pant, and Mani Nepal. 2018. "Costs and Benefits of Implementing REDD+ in Nepal." In *REDD+ in Nepal*, edited by Sindhu Dhungana, Mohan Poudel, and Trishna S. Bhandari. REDD Implementation Centre, MoFE, Government of Nepal. Kathmandu: Jagadamba Press.

Rangan, Haripriya. 2000. *Of Myths and Movements*. London: Verso.

Regmi, Mahesh C. 1967. "Religious and Charitable Land Endowments: Guthi Tenure." *Land Tenure and Taxation in Nepal* IV, no. 12: 1–250.

Robbins, Paul. 2004. *Political Ecology*. Cambridge, MA: Blackwell.

Satyal, Poshendra, Esteve Corbera, Neil Dawson, Hari Dhungana, and Gyanu Maskey. 2019. "Representation and Participation in Formulating Nepal's REDD+ Approach." *Climate Policy* 19, no. 1: 8–22.

Sen, Amartya. 1999. *Development as Freedom*. New York: Anchor Books.

Shah, Saubhagya. 2018. *A Project of Memoreality*. Kathmandu: Himal Books.

Sheppard, Eric, Philip Porter, David Faust, and Richa Nagar. 2009. *A World of Difference*. New York: The Guilford Press.

Sherpa, Pasang D., Tunga B. Rai, and Neil Dawson. 2018. "Indigenous Peoples' Engagement in REDD+ Process: Opportunities and Challenges in Nepal." In *REDD+ in Nepal*, edited by Sindhu Dhungana, Mohan Poudel, and Trishna S. Bhandari. REDD Implementation Centre, MoFE, Government of Nepal. Kathmandu: Jagadamba Press.

Shrestha, Arpan. 2019. "Everything You Need to Know About the Guthi Bill." *The Kathmandu Post*, September 11.

Solnit, Rebecca. 2016. *Hope in the Dark*, 2nd ed. Chicago: Haymarket Books.

Spivak, Gayatri C. 1988. "Can the Subaltern Speak?" In *Marxism and the Interpretation of Culture*, edited by C. Nelson and L. Grossberg. Basingstoke: Palgrave Macmillan.

Stanford, Jim. 2015. *Economics for Everyone*, 2nd ed. London: Pluto Press.

Vizenor, Gerald R. 1999. *Manifest Manners*. Lincoln: University of Nebraska Press.

Wagle, Radha, Soma Pillay, and Wendy Wright. 2020. *Feminist Institutionalism and Gendered Bureaucracies*. London: Palgrave Macmillan.

Waring, Marilyn. 1990. *If Women Counted*. New York: HarperCollins.

Whelpton, John. 2005. *A History of Nepal*. Cambridge: Cambridge University Press.

Wright, Melissa. 2006. "Differences that Matter." In *David Harvey: A Critical Reader*, edited by Noel Castree and Derek Gregory. Malden, MA: Blackwell.

7 A multi-scalar postcolonial political ecology of the commons in an era of climate crisis

As climate change rapidly moves from being an impending crisis to an ongoing planetary emergency, it is timely to critically assess the efforts that have been made by the international community and their reverberations at various sites and scales. I have argued that climate change is a crisis of the commons at multiple sites and scales, riven with struggles to control them for profit, well-being, or both. Since power dynamics are central to the struggles over the commons, political ecology offers a useful conceptual tool for assessing where we stand on our global efforts to address climate change. However, I have found it necessary to stretch the terrain of political ecology both upwards/outwards to the international North/South frame of reference; on the ground to unpack the 'local' or 'regional' context with more granularity than it has been afforded for South Asia; and beyond economic determinism, in order to do justice to understanding more fully the connections between the atmospheric commons and one particular other commons – Nepal's community forests. In the process I have found many confirmations and a few exceptions to commonly held truisms in the critical social sciences about understandings of colonialism and capitalism; about commons-commodity binaries; about modernization, development, and sustainability; and about the struggle for justice – as well as numerous opportunities for connecting theories, processes, and discourses across sites and scales. I have found Gramsci's notion of hegemony to be useful across several contexts and suggest the desirability of exploring the connections between counterhegemonic struggles at multiple scales.

A geographical lens is particularly helpful in elucidating the politics of climate justice, since justice claims are made – and refuted – on the basis of spatial and scalar configurations. In this sense, the contested spatial categories in the context of climate treaty negotiations were North/South, represented by the Annex I/non-Annex I divide, which, after a valiant struggle on part of Global South negotiators, was nixed in the Paris Agreement. Since scale is socially constructed and – I argue – is made real by its institutionalization, the climate negotiations offer a case of how a particular scalar/spatial configuration of global inequality – North/South – has become invalidated, with implications for climate mitigation justice. Another scalar configuration, pertaining to nation-state, is reified – though contested – and continues to strongly shape imaginaries of climate justice, despite the global nature of the problem. I suggest that categories of social difference such as race/Indigeneity/caste, class, and gender also could be seen as scalar categories which may be institutionalized, as in the case of Chelibeti CFUG. As identity categories, they may be conceptualized at various scales – each with their internal core-periphery dynamics – and problems arise when justice claims assume homogeneity or equivalency across space or conflate the particular with the universal, enabling problematic extrapolations across space.

In theorizing a multi-scalar political ecology of climate justice, structures of patriarchy, capitalism, and colonialism should be considered not only in a diffuse and general sense of the global – which is necessary and apposite to the global nature of GHG emissions, although not sufficient – but also at the site of institutions, both in the local-national context and the context of international relations, where the rules and norms of fairness in the relationships that bind the global community together are written, debated, and re-written, as we have seen with the movement from the Kyoto Protocol to the Paris Agreement. In addition to paying attention to the structural relations between core and periphery, we need to pay attention to the discursive politics of recognition and representation, their inherent politics of scale, and their resultant implications for distributive and participatory justice, as well as structural reforms at the site of institutions, since this is where decisions are made about the fate of the commons, local to global. The climate crisis has entrenched pre-existing inequalities at multiple scales, but it has also emboldened marginalized constituents at multiple scales to demand climate justice, understood in a variety of ways. A challenge and an opportunity moving forward is to support disparate efforts to seek justice while holding them accountable to the struggles of others. A successful counterhegemonic effort to destabilize the dominant paradigm of a patriarchal-modernist-capitalist-coloniality requires complementarity across disparate counterhegemonic struggles and for the recognition of a plurality of struggles (Grosfoguel 2011; Laclau 1990) as well as their intersectionalities (Acha 2019; Sultana 2013). Since successful North-South counterhegemonic efforts seem to require the building of counterhegemonies to global coloniality within both the North and the South (Cox 1993), there is a need to scale up the politics of the local commons and to scale down the politics of the global.

A tragedy of the atmospheric commons: Kyoto Protocol to Paris Agreement

What sets the commons apart from an open-access resource is a commonly agreed-upon and enforced set of rules and norms for resource use by a well-defined community. The struggle for community forestry in Nepal and the struggle for a global climate treaty are both struggles over the commons. Many challenges elude the scaling up of a successful resource commons from local to global context, including how to enforce sanctions for bad behavior and exclude free-riders (Ostrom 2014), yet the need for globally agreed upon norms cannot be denied when we cannot escape being part of a global community where the impacts of climate change driven by some have life- and livelihood-threatening effects on others who may have played a negligible part in and who do not benefit from the processes that contributed to it. The Kyoto Protocol's CBDRRC principle was an attempt to establish equitable norms for an otherwise unwieldy and diffuse resource commons. The United States has so far played an unfortunate role of obstructionist and free-rider in the global efforts to curb the climate crisis, since at least as far back as 1992, while simultaneously using its political clout to shape the treaty by inserting the market-based flexibility mechanisms that weakened it. After lobbying to insert a cap-and-trade mechanism in the Kyoto Protocol, the United States infamously refused to ratify it because it was claimed to impart an unfair economic advantage to countries such as India and China that the United States believed should not be given advantages reserved for developing countries. Such US exceptionalism in matters of global environmental responsibility was demonstrated in stark terms in a recent announcement that "Americans should have the right to engage in commercial exploration, recovery, and use of resources in outer space . . . and the United States does not view it as a global commons", announcing a free-for-all race for mineral resources in the moon, Mars, and other celestial bodies (Executive Order 13914 of April 6, 2020, 20381). This kind of blatant disregard for other members of a global community is in large part to blame for the exacerbation of the climate crisis. Global South countries have relentlessly tried to exert pressure on the United States and other Global North countries to accept responsibility and take accountability for their part in exacerbating the climate and other ecological crises. A wide range of environmental treaties exist today as a result of these efforts, and they exist in parallel to numerous international agreements pertaining to world trade. The power dynamics in both sets of negotiations roughly mirror a core-periphery status quo established since European colonialism shaped the current contours of global power. This is why North-South environmental politics in the context of climate treaty negotiations have mirrored the NIEO discourse.

In political ecology and political economy academic circles in the United States, I have encountered a strong tendency to disavow the UN deliberative process to formulate a global climate treaty as an elitist space, and one that has been co-opted and captured by neoliberal capitalism. The Kyoto Protocol was long vilified for being too weak and ineffectual by US activists and academics, and simultaneously by US negotiating teams for being too extreme in its distributive justice intent. The convergence of the two forces – one as key actor in opposing North-South justice claims, and the other's complicity – is uncanny and discomfiting to me as an academic based in the Global North. Global South negotiating teams fought to protect the Kyoto Protocol with its institutionalization of the CBDRRC principle but subsequently were not able to. The Copenhagen Accord successfully dismantled the North-South distinctions built into the Kyoto Protocol, and the Paris Agreement successfully removed the CBDRRC from being an organizing principle for a global climate treaty. Global South actors lost a battle to keep Global North countries accountable for their disproportionate encroachment over the atmospheric commons. What has survived from the transition from the Kyoto Protocol to the Paris Agreement is the weak link of carbon trading and offsetting provisions, introduced by the United States, which now continues in the guise of "results-based payments" and "internationally transferred mitigation outcomes" (United Nations, Paris Agreement 2015, 6–7). The United States, the erstwhile hegemonic power in the world, the one with the most responsibility and capacity – financial and technological, if not moral – and has done the least to accept its global obligations, has no doubt left an indelible mark on the only space we have for the formulation of rules and norms that govern the fate of the global atmospheric commons.

But power is fluid, not static. Power struggles will continue in the course of defining the parameters of the provisions of the Paris Agreement. I imagine that the CBDRRC principle may yet be revived and strengthened, possibly with a more nuanced depiction of 'differentiated responsibilities' than the Kyoto Protocol's binary categories, and with more meaningful representation of the global community than at the level of state representatives. Meanwhile the current focus of North/South negotiations appears to be on operationalizing financial resources for 'losses and damages' for the inevitable consequences of climate change and on determining the terms of trade for carbon. Even as I agree with critics that carbon trading is an ineffectual way to mitigate greenhouse gases, I maintain that it is ethically problematic to emphatically state: "No REDD+" while providing no alternatives to the financial resource flows that would emanate from such a transaction that forest stewards in Nepal are clearly desirous of. As a result of my research in community forestry, my initial understandings of a commons versus commodities framework have evolved to consider how users of a forest commons may wish to sell forest products as commodities while following the rules and norms of governing the forest commons they have established. It is how they make their livelihood. Respecting the agency and dignity of stewards of an existing forest commons, and compensating them for their valuable contributions to climate mitigation, is important, ideally in the form of an ecological debt, but even in the form of carbon finance. Neoliberal capitalism and the commodification of the atmospheric commons are indeed highly concerning, but communities at various peripheries of the patriarchal-capitalist-colonial world order exert their agencies in different ways – some by taking to the streets in protest, others by negotiating power dynamics vis-à-vis the state through existing institutions that have been earned through populist struggle. A 'radical versus reformist' binary lens to evaluate and rank these diverse strategies to deal with neoliberal expansion does not do justice to the multi-faceted struggles on the ground and ignores the importance of complementarity between a 'war of movement' and a 'war of position' from a Gramscian perspective. Absolute caps on emissions would certainly be a more powerful way to reduce GHGs than cap-and-trade, but since the international community has proven unable so far to come to a collective agreement regarding mandatory caps on emissions – nor climate reparations – there needs to be efforts to ensure that the terms of carbon trade are fair at multiple scales, North-South, state-society, and within the community. For this to happen, core-periphery dynamics of international trade must be addressed at the structural level; therefore, there may yet be opportunities for a revitalization of NIEO-esque politics and a counterhegemonic ecological debt agenda within the parameters of the Paris Agreement.

How to subvert a tragedy of the atmospheric commons: take back the other commons?

If the real tragedy of the atmospheric commons is its commodification, and if the likelihood of the international treaty-making space to establish effective and equitable rules and norms for addressing global climate change is slim, a promising way forward may be to focus on other commons that are connected to the atmosphere, whether as a sink or source, or in other forms. Rather than taking the specter of global capitalism head on, in all its bewildering scope and scale, this approach enables conceptualization of many possible interventions, departing from a one-size-fits-all-esque grand theory of anti-capitalist politics. It is also conducive to recognizing already ongoing struggles – and to give credit where it is due – such that the role of the activist-scholar is to bolster these ongoing struggles rather than to dole out generic prescriptions from a superior all-knowing position. The notion of taking back the commons is hardly new. Gibson-Graham et al. (2013) noted many examples of people taking back myriad commons from powerful economic interests: credit unions, food co-ops, the Montreal Protocol to thwart ozone thinning, among many others. Staunch socialists – of the purportedly true kind – have told me that these efforts do not go far enough in confronting capitalism. Perhaps, but these are excellent places to start. We are multi-faceted beings. We live parts of our lives as consumers in a capitalist world and parts of our lives within commons-like spaces. In our individual lives, we can reclaim more of these spaces to the extent that is feasible. We also live in hybrid societies, part capitalist, part something else, since hegemony is never complete. The charge is to seek the commons in the 'something else' and to grow and strengthen those spaces. We might not be able to eradicate capitalism in one fell swoop, but we can work against its dominance in all spheres of life. We may not be able to extract the market, that is, commodification, out completely from the commons – markets have existed long before the beginnings of West-driven capitalism as we know it – but we may be able to help shift power over resources and market transactions from the hands of the few to the hands of the many, establishing norms and rules for how markets should operate. We may not be able to do this in all contexts, but we can certainly look for contexts where these struggles are already organically occurring and build our theories from there; not by appropriating others' struggles and passing them off as *our theories* but by acknowledging where we derive our radical politics.

In this regard, and in light of the urgent task of taking back the commons to avert "dangerous anthropogenic interference of the climate system", the struggles of Ecuadorian civil society to mobilize their government to leave fossil fuels in the ground in the Yasuni Ishpingo-Tambococha-Tiputini region of Amazonia – with inspiration from Nigerian activists pushing to "leave oil in the soil"' – have blazed a path for many in the climate justice movement to adopt 'keep it in the ground' as their mantra (Temper et al. 2013). Related struggles in resisting the Dakota Access Pipeline, the Keystone XL Pipeline, the Cherry Point coal terminal, and hydraulic fracturing projects in the United States among others are all, I suggest, ways to take back the fossil fuel commons, whether to protect Indigenous sovereignty over sacred lands and waters, or for the protection of local ecologies from environmental and health risks, or for the purpose of climate mitigation. Contrasted with other takings back of the commons, these struggles seek to enclose the commons, keeping them out of extractivism and commodification. When we think of the enclosure of the commons, it is typically in the sense of a powerful entity taking the commons out of the reach of commoners. Here it is just the opposite, suggesting that not all enclosures are problematic. In fact, the enclosure of the fossil fuel commons is a necessity for successful amelioration of the climate crisis, as recognized by IPCC scientists (Rockstrom et al. 2014). It is noteworthy that although the proposal for Yasunization was articulated as a critique of capitalism and of carbon trading, it was taken up by the Ecuadorian government in 2007 under the condition of partial compensation – USD 3.6 billion from the international community for lost revenues – in exchange for leaving intact 846 million barrels of oil in the Yasuni region and preventing 410 million tons of CO2 emissions. Its advocates saw the local and global commons as linked in a relationship of mutual obligation: "It is firstly about the right of communities to decide what happens in their territories . . . At the same time, it is a call for international action and co-responsibility for stewardship over the shared global commons . . . to respect Indigenous rights, keep biodiversity intact, and avoid carbon emissions" (Temper et al. 2013, 8). Although this effort would be voided in 2013 when the funds were not forthcoming, there have been other populist struggles to enclose the fossil fuel commons, including divestment strategies focused on the other end of the commodity chain, leading to 'stranded assets' (McKibben 2019).

As with the forest commons in Nepal, the struggle over the commons is not always a struggle between communities and corporations – it is often a struggle between communities, often Indigenous, and the state. This is also true in the case of struggles of the Adivasi tribes in the Indian states of Telangana and Jharkhand, among others, who have launched a prolonged struggle for self-determination over their *Jal, Jangal, Zameen* commons against a repressive state (Damodaran 2006; Roy 2014). The Nepali state has similarly launched multi-pronged assaults on its Adibasi/Janajati populations, a recent iteration of which was the 2019 Guthi Bill, designed to enclose Indigenous lands and institutions and further disenfranchise Newar autonomy over their resource and cultural commons (Shrestha 2019). The origin of commons-oriented forest govern-ance was not an accident but was the outcome of a deliberate intervention to bring about a devolution of power from the center to the margins. Such institutional experiments have been attempted in other contexts with varying degrees of success (Gilmour 2016). While CF as it exists in Nepal may not be replicated everywhere, it may be worth asking whether there are other fertile grounds where such formal land tenure reforms or other forms of devolution over resource governance can be emulated. While the risk of co-optation by neoliberal interests persists, their power should not be overestimated, neither the agency of commoners underestimated. Just as the atmospheric commons is necessarily interlinked with the forest commons, the atmospheric commons is also linked with other spheres of life on earth – fossil fuels, food and agriculture, water resources, technology and knowledge, disaster response – that may be brought into common pool resource governance arrangements. The idea is not to seize power but to share it, such that those who would be affected by decision-making be the ones making decisions about how to use and distribute resources, and how to cope with risks, possibly in a polycentric structure that connects centralized governance at the international and/or nation-state scales with multiple collaborating centers scaled to meaningfully sized communities,

Figure 7.1 Kayaktivist protest against Arctic oil drilling; in the distance, the Pioneer Polar oil rig, Elliot Bay, Seattle, May 2015

depending on the resource to be governed. As Nepal's CF showed, such devolution of power for self-governance does not come about without struggle, and they are not without their flaws, but if equitable commons-like institutions could be created, diffused laterally, and scaled up, a successful counterhegemony against patriarchal–capitalist coloniality may yet be possible.

Implications for climate adaptation

My research over the past decade has focused exclusively on climate mitigation. During this time much of the climate epistemic community has shifted its focus on climate adaptation, for good reason. Even with the best of efforts, perturbances to the climate system are inevitable with the level of GHGs already in the atmosphere (Adger 2001). Manifestations of these changes abound in the sites of my fieldwork – heat waves in Delhi, dengue outbreaks in the Nepali Terai, the melting of glaciers in the Himalaya, floods across South Asia. In August 2017, during my first visit to Chitwan, I had just left Sauraha for Bharatpur after completing fieldwork when the intense monsoon rains created floods so severe in the Rapati River that other guests at the place I had stayed at the night before had to be ferried out of town on elephant back. While my own suffering in this precipitation event was no more than the cancellation of flights out of Bharatpur, the community members in Sauraha and nearby villages suffered a death toll of 70, loss of crops and livestock, and water-borne disease outbreaks (Sharma 2017). Rhinos from Chitwan National Park were swept away in the floods and some lost their lives. In March 2019, months before I returned for a third year of fieldwork, Nepal recorded its first ever tornado to make landfall in the country, in the Chitwan National Park area, killing 28 people and injuring more than 1,100 (Mallapaty 2019). Clearly this is a place where climate adaptation efforts are needed more than climate mitigation efforts, given the already negligible carbon footprints of people here, and the fact that by virtue of community forests and national parks, the region already contributes to carbon sequestration whether or not they are compensated for it. Given the negligible financial gains CFUGs would derive from REDD+ relative to their contributions towards climate mitigation, and the constraints placed on customary use of forest products, the importance of enhancing community resilience as co-benefits – achieved by strengthening, not weakening, self-governance (Gilmour 2016; Poudel, Rana, and Thwaites 2018) – cannot be overstated.

Mainstream climate adaptation efforts have tended to focus exclusively on technological or economic adaptation solutions, whether by building levees or providing temporary flood relief. In these approaches vulnerability is conceptualized as biophysical, resulting from external shocks that communities need protection from. Such an approach may be necessary but not sufficient – the social and structural determinants of vulnerability must be addressed to prepare communities from climate disturbances. As Taylor (2014) articulates, the representation of weather events caused by global warming as 'natural' disasters belies their socially co-produced nature, precluding an understanding of contextual vulnerabilities that are exacerbated by climatic shocks. The violence of climate change is often preceded by forms of slow violence that have in many cases been precipitated by a US-driven project of neoliberalization and militarism (Nixon 2011; Parenti 2011). It follows that a longer-term approach to building resilience for climate adaptation entails enhancing the capabilities of people and communities to be socially, ecologically, and economically resilient against policies that weaken them. Enhancing the self-governance capabilities of communities is key to bouncing back after a setback, as well as to be able to meet basic needs during an emergency, whether in the form of food, water, or shelter. Despite research documenting that humans can self-mobilize to form mutual aid networks to help one another in times of crisis, official disaster responses tend to often be highly centralized (Solnit 2016), and increasingly utilize private contracted parties in an era of disaster capitalism (Klein 2008; Paudel and Le Billon 2020). While moments of crisis can activate both these competing forces (Barrios 2016), expanding the commons is the resilience-building strategy that will help societies emerge stronger from a crisis. While communities may come together organically to respond to moments of crisis, a structural approach to adaptation planning would institutionalize such efforts to empower them with resources and self-determination capabilities.

Building resilience also entails bringing the productive forces for the sustenance of communities within the domain of members of the community for self-reliance. This does not pertain only to resources such as land, water, and food but also to formal institutional arrangements such as CF and Guthi that formalize local governance of these resources (Lekakis, Shakya, and Kostakis 2018). It is also about knowledge, whether critical scientific information and innovations are brought into the public domain for the common good, or the importance of traditional knowledge – local or Indigenous – for coping with unpredictable environments, is validated. Ostrom (2014) noted that adaptive governance is critical for complex natural systems that are prone to uncertainty, particularly in the face of climate change, offering polycentric governance as a way to share power between the state and communities within the jurisdiction of the state. Empowering localized resource users in making adaptive management decisions based on their traditional and local knowledge of their environments acquired over time is not only strategic for distant centers of power where expertise may be more abstract but also for strengthening the political economic capabilities and resilience of communities by ensuring their control over the means of production during an emergency, and beyond. Such polycentric governance can also help facilitate the negotiation between rights and obligations pertaining to climate and environmental change between communities and the state; and scaled up to the international level, can help coordinate them across disparate commons affecting and affected by climate change. At the level of the state, sharing governance power with local communities enables an already overextended state to devolve responsibility and decision-making in ways that promote innovation and experimentation that is context-specific, using or reviving Indigenous/traditional knowledge systems where present, and enables the most marginalized to participate in, benefit from, and shape public spaces and resources (ibid; Nayar 2020; Vivekananda, Schilling, and Smith 2014). The challenges posed by climate change appear to reveal the limits of the neoliberal modernist project in ensuring social and ecological resilience and are in my view necessitating adaptation responses that privilege alternative ways of being, knowing, and governing.

Towards a geography of climate justice

> We don't want purity, but complexity, the relationships of cause and effect, means and end. Our model of the cosmos must be as inexhaustible as the cosmos. . . . It is not the answer we are after, but only how to ask the question.
>
> (Shevek, in Le Guin 1974, 226)

The pursuit of environmental justice in an era of climate change can only be successful if meaningful efforts are made to understand and dismantle the legacies of colonialism. This entails being open to the many variations of colonization and resistance to colonial projects at various sites and scales while refraining from the tendency to conflate particular experiences with colonialism with a universal idea of decolonization. A challenge is to see particular struggles with climate and environmental justice at disparate sites and scales as connected yet distinct. For instance, a logic of dominance and usurpation of resources by certain groups, resulting in the disenfranchisement of others, may be a common pattern, but the particular contexts may be too different to indiscriminately extrapolate the experiences of one context to explain processes in another. Race/Indigeneity, class, gender, the axes of difference that constitute climate justice discourse in the United States are inadequate to comprehensively explain the experiences of distant places such as India and Nepal. Likewise, the practice of taking back the forest commons in Nepal may not serve as a 'model' for taking back the commons elsewhere. I suggest that a necessary element in the pursuit of climate justice everywhere is the reclamation of the commons from the dominance of historical colonial powers, at particular sites as well as at the site of the international. This entails not only a shift in power over control and access to resources but also a challenge to the epistemological dominance of certain scientific-modernist ways of knowing and responding to climate change. Gramsci's notions of hegemony and counterhegemony are useful in understanding how these power shifts are dynamic and contested and how counterhegemonic struggles at one site/scale influence those at others, such that it may be possible the struggles in particular places may not be fully realized unless they are also connected with global counterhegemonic struggles, and vice versa (Cox 1993; Solon 2019). The various ways in which the causes and effects of climate change throw people and countries from disparate parts of the world into a common metaphorical pool further necessitate disparate struggles to be seen as connected and validate the need of polycentric governance of global common pool resources such as the atmosphere.

I am therefore arguing for a counterhegemonic idea of taking back the commons at multiple scales to be understood as decolonization and climate justice. It may look different in different contexts, and the process is possibly partial and may never be complete. The NIEO proposal represents one aspect of an effort to decolonize power relations in international trade. Similar efforts continued in the context of formulating an equitable, fair, and binding global climate treaty, in recognition of the acute vulnerability of the Global South to the impacts of climate change. These were efforts to seek economic and environmental justice by recalibrating the rules of global trade of resource commodities to achieve distributive and participatory justice by making structural changes in the world system of core-periphery power relations. It is possible that these efforts were unsuccessful because they didn't go far enough in challenging the Global North's hegemony over how economic progress is conceptualized and measured. The struggles for localized governance of forests, land, and water in Nepal and India and elsewhere are particular struggles for environmental justice. These localized struggles to take back the commons are not always made by Indigenous groups, but learning from Indigenous struggles is vital because they have the longest and most consistent history of struggles against dispossession for 'primitive accumulation' to remain resilient in the face of colonial threats. As the Newar Guthis in Nepal and the successful struggles of Indigenous Nations and tribes in Washington show, when Indigenous people of particular places succeed in their struggles to take back the commons – whether for reasons of subsistence, sustenance, or cultural survival – everyone else in those places reap the benefits. The historic decommissioning of the Elwha Dam and the Boldt Decision, both of which have only been possible due to struggles of Washington's Native tribes for self-determination and sovereignty, resulted in the return of salmon in the region (Brown 1994; Mauer 2020). As the more successful localized struggles for justice show, decolonization is not only about reclaiming access to and control over resources but also about reclaiming ways of knowing and being that colonialism sought to erase.

Our understandings of climate justice across sites and scales must also make space for complexity. European colonialism and subsequent economic globalization have created patterns of domination that are not possible to uproot in absolute terms. We cannot go back in time and start again anew. Postcolonial struggles have often co-opted the tools of the oppressors in their quest for survival and survivance. As such, many counterhegemonic struggles have been marked by the inability to shed completely the ways of the colonizers, such that struggles against colonial and neocolonial hegemony often occur within the economic, cultural, linguistic, and patriarchal structural constraints that were inherited from colonialism. Sometimes intersecting layers of oppression – for example, a dominated state that itself dominates others – complicates these struggles further. Different groups are therefore at different stages of desire and ability to engage in a true counterhegemonic struggle against patriarchal-capitalist-colonialism. I believe it is important to acknowledge the validity of all struggles – even relative and partial ones – and to not disavow them as failures. The struggle against the privatization of water in the Cochabamba state of Bolivia is a good example to understand the complexities and contradictions involved in a struggle against the forces of a neoliberalized state (Olivera and Lewis 2004). Solon (2019, 261) similarly examines the complexities of decolonization in regions where there is a desire to return to the Indigenous practice of *Vivir Bien*, an alternative paradigm and way of being:

> Vivir Bien is not possible in a single country in the context of a global economy that is capitalist, productivist, extractivist, patriarchal, and anthropocentric. If this vision is to advance and thrive, a key element is its articulation and complementarity with other similar processes in other countries.

Counterhegemonic movements to establish a more ecologically just and humane paradigm spread across space as localized struggles and Global South struggles against the hegemony of the Global North ought to draw on one another for strength and legitimacy.

Race, class, gender, and Indigeneity across space

EJ scholarship has evolved from a predominant focus on the spatial distribution of 'environmental bads' to consider participatory justice and recognition to ultimately consider seriously the structural determinants of distributive disparities and hurdles to equitable and meaningful participation. Structural explanations have focused on race, class, and gender inequalities, consequently offering critiques of institutionalized racism, capitalism, and patriarchal structures, as well as intersectional lens such as racial capitalism and reproductive justice (Pulido 2016; Chiro 2008). Increasingly there is much needed recognition also of settler colonialism and the resultant production of environmental injustices for Indigenous people (Whyte 2018). All of these explanations have broadened the analytical purchase of the EJ lens and are often and increasingly visible in climate justice discourse as well. To see climate justice equated as racial justice or Indigenous sovereignty is not uncommon. While seeing these as positive developments in the discourse on climate change, I wish to argue that we need to broaden our understanding of the structural drivers of the climate crisis beyond an exclusive focus on struggles and discourse relevant largely within the United States, or even the Americas, to seek to understand how the legacies of colonialism have helped create and further exacerbate the climate crisis and vulnerability of people in a wider range of settler colonial and postcolonial contexts. Broadening the scope of climate justice beyond an exclusive focus on the Americas enables the kind of multi-scalar and intersectional counterhegemonic politics that are needed, as well as to appreciate convergences, divergences, complicities, contradictions, productive tensions across contexts, and to recognize that we need to make space for multiple understandings of climate justice.

Omi and Winant's (2015) notion of racial formation is particularly helpful in understanding that people are racialized rather than belonging to self-evident groups labelled as race or caste. Even as these are socially constructed categories, their power in shaping identities appears enduring across geographical space, to create structures of racism in the United States as well as Bahunism in Nepal. But due to the contingent nature of these social constructions, the 'race' lens has limits when it comes to encompassing the majority of the world's population who, despite being non-White, do not respond to a 'race'-focused discourse as much as they do in the context of the United States (Spivak 1988). Racial justice has a particularly powerful meaning in the United States, particularly in a reinvigorated era of BLM in an environment of entrenched institutionalized racism, particularly against Black people. BIPOC in the United States have heightened vulnerability to climate impacts, as the COVID-19 pandemic has demonstrated (Valenzuela, Crosby, and Harrison 2020). Efforts to equate racial justice with climate justice have consequently increased as environmental and climate-focused organizations scramble to show allyship with BLM. The injustice of racial violence against a persecuted group – impossible to unsee due to social media trends and current technology – has mobilized massive movements for racial justice, including in the form of reparations for slavery and settler colonialism. But racial justice in a climate crisis must extend to BIPOC outside the United States, in the Global South, who are facing highly uncertain futures – including loss of life and livelihood – due to climate change precipitated by the Global North. This is no less a form of violence, and its impact is exacerbated by already-occurring patterns of slow violence, also often created by Global North hegemony (Nixon 2011). Further, there is a need to avoid economic determinism in examining struggles for racial/Indigenous/caste justice, gender justice, and North/South justice, that periodically reemerges in the disavowal and minimization of 'identity politics' when these struggles are not couched within an analysis of capitalism (Acha 2019; Estes and Dunbar-Ortiz 2020; Omi and Winant 2015).

Indigeneity is distinct from race as a social marker of difference, but it is often conflated in ways that are facilitated by the notion of racial formation. Therefore even if Indigenous claims owe their validity to being the first known settlers of a place and an established history before the arrival of later settlers with a colonizing mission, in contemporary racialized discourse, they tend to be categorized or perceived in terms of ethnicity, intermingled with later settlers, although an acknowledgment of original settlement and connection with the land is not uncommon. Even if Indigenous histories predate modern conceptions of nation-state, contemporary Indigenous discourse often is strongly shaped by the territorial trap of the reified nation-state (Agnew and Corbridge 1995). Indigenous people in what is today the USA have a special government-to-government relationship with the federal government institutionalized by legal treaties (Brown 1994). Therefore, while a subjugated and disenfranchised group as a whole – categorized as Native Americans despite their particular identities tied to specific places in this continent-sized country – recognized Indigenous tribes and Nations have leverage through treaties with the United States. The politics of Indigeneity in places such as India and Nepal are tremendously different than in the United States. There are no treaties signed between the state and the Indigenous people therein, such that the struggles for self-determination are much more tenuous. Although not often recognized as such, elements of settler colonialism exist in South Asia despite the end of the era of British colonialism – as indicated by the emergence of a Khas/Arya ethnic group in counter-distinction to the Adibasi/Janajati, understood as Indigenous Nationalities. Yet with a history of migration flows within the Eurasian continent, it is not as easy to make clear demarcations between Indigenous and settler colonial identities in South Asia. There has been a process of Indigenous identity formation akin to the process of racial formation that is contextually contingent and intermingled with caste and ethnicity. Indigenous groups in South Asia may have as much legitimacy as those in the United States, but the absence of formal legal arrangements means that these claims cannot be manifested in ways that they do in the United States.

Colonialism and imperialism, hegemony and hubris

Britain was hegemonic in the world during the colonial era, and the United States has been hegemonic in the post-colonial era, even as the post-colonial project was never completed, and the United States is recognized as a settler colonial state. As the emergence of economies of countries such as China threaten a move to a multi-polar world order, the United States has shown a desire to remain the world hegemon, in part through contemporary struggles over the atmospheric commons over the past three decades (Ziser and Sze 2007). The hegemony of US imperialism as a nation-state, as well as a key neoliberal-capitalist force in the world, is inherently a driver of the climate crisis, even as its obstructionism towards a global climate treaty has impeded efforts towards a coherent response to the crisis. While this is clearly established, what is less clear are the relations between those who are subjected to US imperialism, colonialism, and racism within the United States and abroad. I believe complementarity and mutual support and learning across disparate counter-hegemonic struggles are important for the hegemony of a US-led patriarchal-capitalist-coloniality to be destabilized, and I offer here some observations on constraints and opportunities for meaningful collaboration.

Supremacy is a common theme that cuts across geographic contexts, manifesting as: White supremacy at large, US exceptionalism in North-South climate negotiations, Hindu nationalism in India, White nationalism in the United States, Bahunism in Nepal, and so on. Another word for supremacy is 'hegemony', but White supremacy conjures up the racialized nature of hegemony quite clearly. I believe White supremacy/nationalism and US exceptionalism combine to entrench power relations between those that are privileged/othered within the United States and those outside of it. This shows up in different ways. The hegemony of a US-centered Indigenous politics over the diverse experiences of Indigenous people in the Third World is one element. The invalidation of North-South justice claims on account of the compromised authenticity of North-South justice claims due to perceived elite Third Worldism and insistence on a class-based analysis and the minimization of Third World identity politics are also forms of White privilege and supremacy. Both have the effect of freezing the North-South status quo that perpetuates the hegemony of the United States. US exceptionalism is also reflected through academia where the hubris of the world's hegemonic nation-state is reproduced and propagated through its engines of knowledge production. Yet here insights from Critical Race Theory – such as the importance of impact vis-à-vis intention, and the notion of color-blindness – are helpful. Just as presumptions of color blindness have been used to invalidate racial justice claims in the United States, North-South inequities have been invisibilized by the domination of 'one-worldism' in climate negotiations to negate claims for North-South distributive justice. The overwhelming invisibility of Third World peoples and contexts in environmental and climate justice discourses – including those of reparations – in the United States is another form of White supremacy and US exceptionalism. The desire and impulse to center the United States in climate conversations and invisibilize the Global South when the global atmospheric commons are implicated is simply not acceptable, particularly in climate justice discourses, where the imperative is to lift the most vulnerable, the most disadvantaged, and the least privileged.

Beyond invisibilization, there are also implicit forms of racism towards the Third World that have shown up occasionally, for instance, when disenfranchised groups in the aftermath of Hurricane Katrina articulated their circumstances as unacceptable because they were relegated to Third World status within the United States, the implication being that those circumstances are okay for distant others: "Katrina put a light on race, class, and indifference. People were treated like we live in the third world, and this is the richest country in the world" (White 2012, 163). Notwithstanding Prashad (2014), the Third World is not *only a project*, it is also a place. It is a place whose people are often deemed 'third class' citizens by people in the First World. In my own teaching, I have occasionally heard from students, disaffection of the notion that the United States has a climate debt to the Global South, as that would take resources away from disenfranchised populations within the United States where its real obligations supposedly lie. These sentiments work in tandem with perceptions of the Third World elite who would siphon away these resources from the most disadvantaged. Along these lines the claim to sovereignty of Third World states such as India is mocked as elitist because it is couched in the language of a right to modernist development. There is a certain hypocrisy in idolizing Indigenous sovereignty within the United States – even as Indigenous Nations themselves are hierarchical and heterogeneous – while invalidating the sovereignty claims of another, formerly colonized, country. Inherent in these double standards is an Orientalist view of Indigenous people as the 'Ecological Indian' and Third World countries as closer to nature or 'backward' and their association with modernity and economic power a contradiction in terms (Ray 2013). Such 'green Orientalism' does a disservice to Native American groups such as the Makah, who have struggled to resume the ancient practice of whaling using modern technologies (Kim 2015; Lohmann 1993), and it certainly does not do justice to India with its political and economic aspirations. I have argued that India should be held accountable to its Adivasi communities, as well as to its neighbors in South Asia, but using this as an excuse to invalidate their North-South equity claims is not warranted. Such Orientalism is a form of racism towards the Global South, and a way to maintain White supremacy; and it is not only White Americans who exhibit these tendencies. As a result, the treatment of the Third World as sacrifice zones for global capitalism – exemplified by the much-discussed leaked memo by Lawrence Summers – appears to be buttressed by a widespread culture of US exceptionalism (Nixon 2011), even as the notion of sacrifice zones within the United States in Native and Black spaces has been problematized and internalized by those who advocate for climate justice.

Reflexivity, complicity, positionality

Spivak (2010) has warned against the dangers of academic complicity with prevailing power geometries. Others have called for greater reflexivity among those who benefit from the privileges of the Western world and its modern rationalities and technologies to avoid complicity in perpetuating various forms of climate injustice (e.g. Salleh 2011; Roos and Hunt 2010). It behooves scholars, activists, and well-meaning citizens in the Global North concerned about complicity to be conscious of their positionality – how they are situated in the world in relation to fossil-fueled empires, and the associated power geometries of climate privilege and climate vulnerability – and how they benefit from the status quo. Any analysis of the climate crisis and solutions need to be situated in this positionality; even if not explicitly stated as such, they need to be understood as situated analyses and prescriptions, to avoid what Haraway (1998) calls the God-trick effect. This is one way to work against hypocrisy in what we say and what we do, as well as to broaden our faculties to produce nuanced prescriptions. Postcolonial theory offers multiple avenues to think of North-South climate justice in nuanced and multi-faceted ways, including how countries such as India and China continue to be subjected to Orientalistic othering practices in climate discourse in the West, how strategic essentialism and hybridity enable a postcolonial politics that assists colonized groups to use the colonizers' tools in challenging an unequal status quo, and how to think reflexively about the associations between critique and complicity in maintenance of this status quo that are being challenged in climate politics. I offer a few observations of occasions where reflexivity is warranted in climate justice discourse.

Equating the climate crisis with modernist and capitalist development for the pursuit of economic growth is de riguer in climate justice discourse within political ecology. Even as it is important to problematize the dominance of modernist science and technology as well as neoliberal modes of development and economic growth over the precolonial/premodern variants, I believe those who benefit from these structures of modernism cannot invalidate the desires of others who seek to do so. As some have astutely pointed out that ecological collapse would be the inevitable outcome (e.g. Sachs 2002), it strikes me as necessary to model a delinking between well-being and modernist economic growth in our own spheres before we proceed to tell distant others what (not) to do. Scholars are prone to adopting a self-identified higher moral ground, for genuine or performative reasons. Practicing reflexivity may dissuade us from imposing idealistic solutions on distant others based on our own romanticized and purist views of identity, well-being, the commons, and so on that we might not choose for our own lives. I believe absence of reflexivity engenders the unconscious process through which the position of the White savior is sustained. It is saviorism to wax poetic over the plight of climate refugees and to support immigrant rights in the United States, while refusing to recognize the economic justice claims and economic growth aspirations of countries that climate refugees may hail from, oblivious to the role played by US hegemony in international relations in precipitating their displacement by undermining the economic capabilities of their countries. A performative advocacy for climate refugees seems easier than conceptualizing Third World countries as being robust economies from where people would not have to emigrate in search of a better life. Migration patterns are, after all, an indication of people's desires, aspirations, and/or desperation.

Just as respect for colonized people's appropriation of the colonizers' tools in their struggle for equality – Bhabha's notion of hybridity – is warranted when considering the legitimacy of Indigenous struggles for self-determination in relation to using 'Western' modes of science, development, and modernization in spaces deemed traditional, there is also a need to recognize the importance of the role of the modern nation-state. Despite claims of a territorial trap and the subsequent irrelevance of the state in a climate of rampant economic globalization and the specter of capital (Agnew and Corbridge 1995), the nation-state continues to be relevant in numerous ways, as the global pandemic has demonstrated. A nation is not just an imagined community; although it is socially constructed – like race and Indigeneity – it is made real by constitutions promulgated, treaties signed, laws written, and passports issued, among many ways in which the relationships between a state and its citizens are congealed. Even as states can be hegemonic towards their people, they also protect them from external threats, as seen in the case of the creation of the nation-state of Nepal in 1768/9. The struggles between the state and its people cannot be used as an excuse by a more powerful outside force to delegitimize the sovereignty claims of a nation-state. In the context of global inequality and global environmental governance, the state appears to be an arbiter between foreign interests and its citizens, working to defend economic equity in international relations, as the role of India in climate negotiations suggests. Countries such as China and India have historically exercised protectionism – and have been chastised for doing so by adherents of free trade – and US-led neoliberalism and structural adjustment policies have weakened this function (Klein 2008; Nixon 2011). State-society dynamics are complex and the state cannot be seen solely as an oppressor – although this may well be one of its functions. Transnational justice movements, say, against large dams in India that disenfranchise Indigenous and other marginalized groups, in the long run still need to work within the parameters of the state, long after activist allies from outside have left. Just as it is important to respect the sovereignty of states, it is important to respect international treaties. The disavowal of the international treaty-making process as elitist by US scholar-activists even as treaties between the United States and Native Nations and tribes are upheld as sacred is yet another (subconscious) strategy of White supremacy through implicit US-centrism – especially prescient when you take note of how the United States, the world's hegemonic power, has yet to sign treaties such as the Basel Convention, whose goal is to prevent the export of hazardous waste to Third World countries. International environmental treaties such as the Kyoto Protocol sought to institutionalize equitable burden-sharing for ecological repair and seek to hold the Global North accountable to its historical responsibility for climate change. The United States single-handedly negated those efforts. Those who enjoy privileges afforded by the world's hegemonic power are complicit and accountable to such US exceptionalism.

Politics of scale: building counterhegemonies across space

Pursuits of climate justice are complicated by complex layered geographies of global inequality. Seemingly straightforward and reasonable prescriptions such as the 'polluter pays principle' or 'per capita equity' are convoluted by the ways in which social categories have formed to defy the easy identification of who exactly the culprits are in particular understandings of climate justice – there are so many – and who exactly they are accountable to. Justice claims conjure up particular identity categories corresponding to particular spatial and scalar configurations, and there is no global consensus over which ones are the ones to follow. That we live in a world rife with global inequalities is uncontested, but how should we operationalize equity? Do we stay with the scale of nation-states – entrusting each nation-state to parse out the internal equity work and power dynamics on their own – or seek out a transnational framework because a state-centric approach falls prey to the territorial trap and may privilege the elite in each nation-state? A sense of difference between the Global North and Global South is generally accepted but challenged due to their state-centrism, heterogeneity, and hierarchies within. How/why to arrive at universal notions of racial/Indigenous, gender, and economic justice in a world of cultural and contextual differences across and within the Global North and South? I have argued for an approach similar to Schlosberg's (2009) that emphasizes pluralism, where different approaches to conceptualizing and operationalizing climate justice do not need to compete for dominance, and where each group's struggle is regarded as valid by others. A successful counterhegemonic struggle may well require disparate struggles to join forces, but this is not likely when one does not respect the other's struggle or trivializes it as, say, identity politics. The way in which BLM has reverberated across the world to double down against racism against Black people in all forms as well as to reinvigorate movements more specific to places – as evidenced by Dalit Lives Matter in South Asia – is indeed a hopeful sign. Indigenous struggles against settler colonialism in the United States – while not extending to problematizing US imperialism and hegemony abroad – offers helpful impetus for conceptualizing patterns of settler colonialism in contexts such as Nepal and India. Whether these grounded movements for marginalized lives in disparate places might also spur structural changes in core-periphery relations such that Black/Dalit/Indigenous lives can be freed from structures of slow violence beset by colonialism is an unanswered question. The 2020 Climate Conference COP-26 was postponed due to the pandemic, but it will be interesting to see what impact BLM – as a counterhegemonic movement that originated in the United States – will have on future UN climate deliberations and prospects for racial justice along the North-South divide.

Meanwhile, layers of unequal power dynamics at multiple sites and scales appear to make possible a wide variety of justice claims, some of which, such as India's claim to atmospheric space based on the Global South's 'right to development', has been problematized. All justice claims that engage in a scale politics are subject to accountability. Constituents of a state can in theory hold the state accountable to the claims it makes at the international scale, as the *GreenPeace India 2007* report did. One of the contradictions of colonialism and the layers of inequality it produces is that marginalized people and groups can 'jump scale' beyond the nation-state to seek justice within. Sometimes struggles at one scale are enabled in complicated ways by disparities at another scale: conversion to Christianity – rather than Marxism – has been an avenue for Dalit liberation from the 'tyranny of caste' for many in Nepal (Pariyar 2019), suggesting colonialism's complex entanglements. Without my mastery of the Nepali and English languages – languages of colonization – I would not have been able to jump scale to overcome the patriarchal, caste, and class constraints to my pursuit of freedom and fulfilment as a human being. When you are at the bottom rung of multiple hierarchies, you can jump scale and claim strategic essentialisms to make particular claims to justice. This is one upside of embodying multiple intersecting oppressions, and it speaks to the agency of those perceived to be the oppressed. External interventions can be critical in addressing power relations in the 'traditional' society that would otherwise remain ignored. In this sense, gender issues in patriarchal society and Indigenous rights in colonial contexts can be addressed with impetus from outside, such as by calling on UNDRIP (Tilsen 2020). As the REDD+/community forestry case study revealed, advocates of gender equality, Indigenous rights, and decentralization struggles in Nepal see Western intervention as a catalyst to further their efforts. Even as universalized Western discourses of feminism and Indigenous sovereignty might be limiting in the context of Nepal, they are often appropriated and adapted by actors to suit their particular contexts. I see this as a scale-jumping exercise where marginalized groups reach beyond the domains of their community and state boundaries and adopt a globalized discourse to achieve specific aims. Neither the Global South nor the Global North is a monolith, and alliance-building across North and South has promising possibilities for the liberation of all marginalized peoples (Pellow 2007). North-South justice is necessary but not sufficient. Well-intentioned Global North members – and global treaties formulated in seemingly elitist spaces – can have a positive role to play in assisting in the ongoing struggles in the Global South. The charge is to intervene in ways that are not paternalistic, patronizing, appropriating, or tokenizing and that do not perpetuate colonial power dynamics or worsen community relations; the challenge is to be reflexive without becoming paralyzed. Respecting the agency of groups one is trying to support is paramount.

Geography has come a long way as a discipline whose role early on was to advance the colonial project through cartography. Today it is still an overwhelmingly White discipline, and yet it offers analytical tools to help make sense of the complexities of the struggle for climate justice. Geographers study nature-society relations across space in a world of difference. As such it is an ideal discipline for helping overcome US-centrism in our understandings of climate justice, as well as to facilitate a much-needed multi-scalar understanding of the power relations that mediate the experience of climate change (Sultana 2013). The geographical emphasis on place can help ground conceptualizations of the atmospheric commons to particular places, enabling an understanding of how disparate commons are connected across scale and space and how many localized and place-based commons are linked to the atmospheric commons conceptualized at the global scale. Climate justice claims have an inherently geographical component to them. Since scale and space are socially configured, individuals exercise their agency in scale jumping to mobilize justice claims corresponding to particular configurations and strategic essentialisms. Contestations over such scale jumping, as well as corresponding spatial and scalar categories, are an important aspect of struggles for environmental and climate justice (Walker 2012). Relatedly, geographical thinking is helpful in making sense of the politics of identity. Identity is not a given; it is forged through struggles over the commons. Unlike other scalar identity categories, the national scale has been reified in the nation-state as an institution. It is at this scale that rule-making for the governance of national resource commons takes place. While the primacy of this identity has been questioned by geographers, it is not yet clear what spatial/scalar configuration is most ideal for resource governance, although ideas of bioregion and watershed seem compelling. To the extent that nation-states continue to be relevant stakeholders in addressing the problem of climate change at the international scale, I suggest that Ostrom's (2014) notion of polycentric governance is a compelling institutional framework for helping bridge the gap between disparate resource users, emitters, and possible subjects of climate consequences everywhere on the one hand and decision-makers, rule-makers, and negotiators at the international scale of climate governance on the other.

The identity categories of race/Indigeneity/caste/ethnicity, class, and gender can be conceptualized and institutionalized from local to global scales. The process of reification is uneven and is constantly contested, conflated, and extrapolated, thus competing claims for climate justice – an idea whose meaning has not been fixed and is thus a floating signifier – struggle for legitimacy, visibility, and dominance. I have argued that since the causes and effects of climate change are globally dispersed – whether or not our mechanisms for accountability are global yet – we cannot continue to promote a climate justice discourse that invisibilizes the circumstances, experiences, and voices of globally marginalized groups by conflating particularistic US-centered climate justice imaginaries to the global scale by assuming their universality. The climate crisis demands more from US-based scholars, policy-makers, and activists.

References

Acha, Majandra R. 2019. "Climate Justice Must Be Anti-Patriarchal, or It Will Not Be Systemic." In *Climate Futures*, edited by Kum-Kum Bhavnani, John Foran, Priya A. Kurian, and Debashish Munshi. London: Zed Books.

Adger, W. Neil. 2001. "Scales of Governance and Environmental Justice for Adaptation and Mitigation of Climate Change." *Journal of International Development* 13: 921–31.

Agnew, John, and Stuart Corbridge. 1995. "The Territorial Trap." In *Mastering Space*, edited by John Agnew. London: Routledge.

Barrios, Roberto E. 2016. "Expert Knowledge and the Ethnography of Disaster Reconstruction." In *Contextualizing Disaster*, edited by Gregory V. Button and Mark Schuller. New York: Berghahn Books.

Brown, Jovana J. 1994. "Treaty Rights: Twenty Years After the Boldt Decision." *Wicazo Sa Review* 10, no. 2: 1–16.

Chiro, Giovanni Di. 2008. "Living Environmentalisms: Coalition Politics, Social Reproduction, and Environmental Justice." *Environmental Politics* 17, no. 2: 276–98.

Cox, Robert W. 1993. "Gramsci, Hegemony and International Relations: An Essay in Method." In *Gramsci, Historical Materialism and International Relations*, edited by Stephen Gill. Cambridge: Cambridge University Press.

Damodaran, Vinita. 2006. "Indigenous Forests." In *Ecological Nationalisms*, edited by Gunnel Cederlof and K. Sivaramakrishnan. Seattle: University of Washington Press.

Estes, Nick, and Roxanne Dunbar-Ortiz. 2020. "Examining the Wreckage." *Monthly Review*, July 1.

Executive Order 13914 of April 6. 2020. "Encouraging International Support for the Recovery and Use of Space Resources." *Code of Federal Regulations* 20381–82, title 3.

Gibson-Graham, J. K., Jenny Cameron, and Stephen Healy. 2013. *Take Back the Economy*. Minneapolis: University of Minnesota Press.

Gilmour, Don. 2016. *Forty Years of Community-Based Forestry*. FAO Forestry Paper 176. Rome: Food and Agriculture Organization of the United Nations.

Grosfoguel, Ramon. 2011. "Decolonizing Post-Colonial Studies and Paradigms of Political-Economy: Transmodernity, Decolonial Thinking, and Global Coloniality." *Transmodernity: Journal of Peripheral Cultural Production of the Luso-Hispanic World* 1, no. 1: 1–38.

Haraway, Donna J. 1998. "Situated Knowledges: The Science Question in Feminism and the Privilege of Partial Perspective." *Feminist Studies* 14, no. 3: 575–99.

Kim, Claire Jean. 2015. *Dangerous Crossings*. New York: Cambridge University Press.

Klein, Naomi. 2008. *The Shock Doctrine*. New York: Picador.

Laclau, Ernesto. 1990. *New Reflections on the Revolution of Our Time*. New York: Verso.

Le Guin, Ursula K. 1974. *The Dispossessed*. New York: HarperVoyager.

Lekakis, Stelios, Shobhit Shakya, and Vasilis Kostakis. 2018. "Bringing the Community Back: A Case Study of the Post-Earthquake Heritage Restoration in Kathmandu Valley." *Sustainability* 10, no. 8: 2798–815.

Lohmann, Larry. 1993. "Green Orientalism." *Ecologist* 23, no. 6: 202–4.

Mallapaty, Smriti. 2019. "Nepali Scientists Record Country's First Tornado." *Nature*, April 12.

Mauer, K. Whitney. 2020. "Undamming the Elwha River." *Contexts* 19, no. 3: 34–39.

McKibben, Bill. 2019. "Divestment Works – and One Huge Bank Can Lead The Way." *The Guardian*, October 13.

Nayar, Varun. 2020. "The Fight for the Commons." *Himal Southasian*, May 25.

Nixon, Rob. 2011. *Slow Violence and the Environmentalism of the Poor*. Cambridge, MA: Harvard University Press.

Olivera, Oscar, and Tom Lewis. 2004. *Cochabamba!*. Cambridge: South End Press.

Omi, Michael, and Howard Winant. 2015. *Racial Formation in the United States*, 3rd ed. New York: Routledge.

Ostrom, Elinor. 2014. "A Polycentric Approach for Coping with Climate Change." *Annals of Economics and Finance* 15, no. 1: 97–134.

Parenti, Christian. 2011. *Tropic of Chaos*. New York: Nation Books.

Pariyar, Mitra. 2019. "Fighting Caste Through Religious Conversion: Early Experiences of Nepali Dalit Christians." The Annual Kathmandu Conference on Nepal and the Himalaya, Kathmandu, July 24.

Paudel, Dinesh, and Philippe Le Billon. 2020. "Geo-Logics of Power: Disaster Capitalism, Himalayan Materialities, and the Geopolitical Economy of Reconstruction in Post-Earthquake Nepal." *Geopolitics* 25, no. 4: 838–66.

Pellow, David N. 2007. *Resisting Global Toxics*. Cambridge, MA: The MIT Press.

Poudel, Mohan, Eak Rana, and Richard Thwaites. 2018. "REDD+ and Community Forestry in Nepal." In *Community Forestry in Nepal*, edited by Richard Thwaites, Robert Fisher, and Mohan Poudel. London: Routledge.

Prashad, Vijay. 2014. *The Poorer Nations*. New York: Verso.

Pulido, Laura. 2016. "Flint, Environmental Racism, and Racial Capitalism." *Capitalism Nature Socialism* 27, no. 3: 1–16.

Ray, Sarah J. 2013. *The Ecological Other*. Tucson: University of Arizona Press.

Rockstrom, Johan et al. 2014. "Climate Change: The Necessary, the Possible and the Desirable Earth League Climate Statement on the Implications for Climate Policy from the 5th IPCC Assessment." *Earth's Future* 2: 606–11.

Roos, Bonnie, and Alex Hunt. 2010. *Postcolonial Green*. Charlottesville: University of Virginia Press.

Roy, Arundhati. 2014. *Capitalism*. Chicago: Haymarket Books.

Sachs, Wolfgang. 2002. "Ecology, Justice, and the End of Development." In *Environmental Justice*, edited by John Byrne, Leigh Glover, and Cecilia Martinez. New Brunswick: Transaction Publishers.

Salleh, Ariel. 2011. "Cancun and After: A Sociology of Climate Change." *Arena* 110: 24–31.

Schlosberg, David. 2009. *Defining Environmental Justice*. New York: Oxford University Press.

Sharma, Gopal. 2017. "Elephants Help Rescue Hundreds From Flooded Nepali Safari Park." *Reuters*, August 13.

Shrestha, Arpan. 2019. "Everything You Need to Know About the Guthi Bill." *The Kathmandu Post*, September 11.

Solnit, Rebecca. 2016. *Hope in the Dark*, 2nd ed. Chicago: Haymarket Books.

Solon, Pablo. 2019. "Is Vivir Bien Possible? Candid Thoughts About Systemic Alternatives." In *Climate Futures*, edited by Kum-Kum Bhavnani, John Foran, Priya A. Kurian, and Debashish Munshi, 253–62. London: Zed Books.

Spivak, Gayatri C. 1988. "Can the Subaltern Speak?" In *Marxism and the Interpretation of Culture*, edited by C. Nelson and L. Grossberg. Basingstoke: Palgrave Macmillan.

———. 2010. "In Response: Looking Back, Looking Forward." In *Can the Subaltern Speak?*, edited by R. C. Morris. New York: Columbia University Press.

Sultana, Farhana. 2013. "Gendering Climate Change: Geographical Insights." *Professional Geographer* 66, no. 3: 372–81.

Taylor, Marcus. 2014. *The Political Ecology of Climate Change Adaptation*. New York: Routledge.

Temper, Leah et al. 2013. "Towards a Post-Oil Civilization: Yasunization and Other Initiatives to Leave Fossil Fuels in the Soil." *EJOLT Report* 6: 1–204.

Tilsen, Nick. 2020. "Opinion: I Stood Up for Indigenous Rights at Mount Rushmore. Now I'm Facing 17 Years." *Newsweek*, August 21.

United Nations. 2015. *Paris Agreement*, chapter XXVII, Environment, title 7d. Paris: United Nations, December 12.

Valenzuela, Jessica, Lori E. Crosby, and Roger R. Harrison. 2020. "Commentary: Reflections on the COVID-19 Pandemic and Health Disparities in Pediatric Psychology." *Journal of Pediatric Psychology* 45, no. 8: 839–41.

Vivekananda, Janani, Janpeter Schilling, and Dan Smith. 2014. "Understanding Resilience in Climate Change and Conflict Affected Regions of Nepal." *Geopolitics* 19, no. 4: 911–36.

Walker, Gordon. 2012. *Environmental Justice*. New York: Routledge.

White, Mia Chalene. 2012. "Gender, Race, and Place Attachment: The Recovery of a Historic Neighborhood in Coastal Mississippi." In *The Women of Katrina*, edited by Emmanuel David and Elaine Enarson. Nashville: Vanderbilt University Press.

Whyte, Kyle P. 2018. "Settler Colonialism, Ecology, and Environmental Justice." *Environment & Society* 9, no. 1: 125–44.

Ziser, Michael, and Julie Sze. 2007. "Climate Change, Environmental Aesthetics, and Global Environmental Justice Cultural Studies." *Discourse* 29, no. 2–3: 384–410.

Index

Printed in the United States
By Bookmasters